Lecture Notes
in Control and Information Sciences 332

Editors: M. Thoma · M. Morari

Shengyuan Xu · James Lam

Robust Control and Filtering of Singular Systems

 Springer

Authors

Professor Shengyuan Xu

Department of Automation
Nanjing University of Science and Technology
Nanjing 210094
P. R. China
syxu02@yahoo.com.cn
syxucn@mail.njust.edu.cn

Professor James Lam

Department of Mechanical Engineering
The University of Hong Kong
7/F Haking Wong Building, Pokfulam Road
Hong Kong
P. R. China
james.lam@hku.hk

ISSN 0170-8643

ISBN-10   3-540-32797-5 **Springer Berlin Heidelberg New York**
ISBN-13   978-3-540-32797-4 **Springer Berlin Heidelberg New York**

Library of Congress Control Number: 2006921744

Springer is a part of Springer Science+Business Media

springer.com

Typesetting: Data conversion by authors.
Final processing by PTP-Berlin Protago-TEX-Production GmbH, Germany
Cover-Design: design & production GmbH, Heidelberg
Printed on acid-free paper   89/3141/Yu - 5 4 3 2 1 0

To A'jin, Fagen, and Junwei

Shengyuan Xu

To Yeeman, Joanie, and Jamie

James Lam

Preface

Singular systems have been widely studied in the past two decades due to their extensive applications in modelling and control of electrical circuits, power systems, economics and other areas. A great number of fundamental results based on the theory of state-space systems have been successfully extended to singular systems. Interest has grown recently in the stability analysis and control of singular systems with parameter uncertainties due to their frequent presence in dynamic systems, which are often the causes for instability and poor performance of control systems. It is known that the control of uncertain singular systems is much more complicated than that of state-space systems because controllers must be designed so that the closed-loop system is not only robustly stable, but also regular and impulse-free (in the continuous case) or causal (in the discrete case), while the latter two issues do not arise in the state-space case. Many recent advances in this area have made use of the matrix inequality machinery, which has become a somewhat standard approach to deriving appropriate mathematical conditions. This monograph aims to present up-to-date research developments and references on robust control and filtering of uncertain singular systems in a unified matrix inequality setting. It provides a coherent approach to studying control and filtering problems as extensions of state-space systems without the commonly used slow-fast decomposition. It contains valuable reference material for researchers wishing to explore the area of singular systems, and its contents are also suitable for a one-semester graduate course.

The topics covered in this monograph are given in parallel for continuous and discrete singular systems. We first present necessary and sufficient conditions for a singular system to be regular, impulse-free (for continuous singular systems) or causal (for discrete singular systems) in terms of linear matrix inequalities (LMIs). By employing these fundamental results, the concepts of generalized quadratic stability and generalized quadratic stabilization for uncertain singular systems are introduced and both robust stability and stabilization results are obtained via these two notions. Versions of the bounded real lemma are then established and solutions to the problems of robust H_∞ control and filtering are derived. The problem of guaranteed cost control is solved via an LMI approach for both continuous and discrete singular systems. Furthermore, necessary and sufficient conditions for the solvability of the positive real control problem are obtained based on the proposed versions of the positive real lemma. When time delays appear in a singular system, sufficient conditions on the solvability of the robust control and H_∞ filtering of singular delay systems are given. For Markovian jump singular systems, the problems of guaranteed cost control and H_∞ control are addressed and state feedback controllers are synthesized. In the continuous case, the LMI approach is developed in this monograph to design desired controllers and filters, which involves no decomposition of the original systems.

The work was partially supported by the Hong Kong Research Grant Council under grant HKU 7029/05P, the National Natural Science Foundation of P. R. China under Grant 60304001, the Program for New Century Excellent Talents in University (No. NCET-04-0508), the Foundation for the Author of National Excellent Doctoral Dissertation of P. R. China under Grant 200240, and the Fok Ying Tung Education Foundation under Grant 91061. Part of the work was completed while the first author was on leave visiting the Department of Mechanical Engineering, The University of Hong Kong.

Nanjing University of Science and Technology, China   *Shengyuan Xu*

The University of Hong Kong, Hong Kong   *James Lam*

December 2005

Contents

1

Introduction and Preview

1.1 Introduction

The classical state-space model for a proper system, corresponding to a set of ordinary differential equations (ODEs), can be obtained by selecting a minimal set of variables (known as state variables). However, there are practical situations where physical variables (referred to as descriptor variables) cannot be chosen as state variables in a natural way to provide a mathematical model in a state-space form. Usually, descriptor variables chosen naturally are not minimal in the sense that they may be related algebraically. This may result in a singular system model since some of the relationships of these variables are dynamic while others purely static. It should also be noted that a state-space model is obtained under the assumption that the plant under consideration is governed by the causality principle, but in certain situations the state of a system in the past may depend on its state in the future. This anticipatory characteristic of a system leads to a violation of the causality assumption. In this case, a singular system model is necessary for the description of such a system. Mathematically speaking, a singular system model is formulated as a set of coupled differential and algebraic equations, which include information on the static as well as dynamic constraints of a real plant. Such systems are called singular systems [20, 21, 36, 108]. In the literature, singular systems are also referred to as descriptor systems [107, 121, 122], differential-algebraic equation systems [19, 96, 97], generalized state-space systems (or generalized systems) [10, 165, 171], implicit systems [6, 7], and semistate systems [131, 132, 145]. They do not belong to the class of ODEs since an ODE does not have any underlying algebraic constraints on its variables. The singular system form contains the state-space form as a special case and thus can represent a much wider class of systems than its state-space counterpart. Applications of singular systems can be found in aircraft modelling [152], chemical processes [95], circuit systems [131, 132], economic systems [120], large-scale interconnected systems [121, 151], mechanical engineering systems [75], power systems [146], and robotics [126].

Singular systems are fundamentally different from state-space systems. The following features in singular systems are not usually found in state-space systems [12, 165, 195]:

- The transfer function of a singular system may not be strictly proper.

- For an arbitrary finite initial condition, the time response of a singular system may exhibit impulsive or non-causal behavior along with the derivatives of these impulses.

- A singular system usually contains three kinds of modes: finite dynamic modes, infinite dynamic modes and nondynamic modes; the undesired impulsive behavior in a singular system can be generated by infinite dynamic modes.

- Even if a singular system is impulse-free, it can still have initial finite discontinuities due to inconsistent initial conditions.

As singular systems constitute an important class of systems of both theoretical and practical significance, they have been the subject of extensive research during the past decades. A great number of fundamental concepts and results based on the theory of state-space systems have been successfully extended to singular systems, including:

- solvability, controllability and observability [1, 23, 27, 32, 58, 78, 92, 172, 215];

- canonical forms [22, 61, 62, 74, 102, 162];

- minimal realizations [33, 47, 67, 76, 94, 128];

- system equivalence [9, 73, 77, 165, 212];

- regularity and regularization [16, 17, 18, 28, 29, 51, 99, 136, 169];

- stability and stabilization [59, 87, 104, 117, 137, 148, 159, 164, 189, 206];

- pole assignment [2, 30, 43, 63, 64, 91, 163, 197, 208];

- linear-quadratic optimal control [12, 31, 98, 100, 138, 168];

- observer design [34, 37, 56, 79, 89, 127, 129, 140, 141, 194];

- Lyapunov theorems and equations [11, 82, 154, 158, 200, 204, 207];

- model reduction [176, 180, 205];

- H_2 control [83, 155, 213].

Other topics related to singular systems such as analysis and control of nonlinear singular systems have also been studied [26, 95, 118, 119, 150, 214]. The results in these studies were obtained under the assumption that the singular systems are free of uncertainties. However, modelling errors and system uncertainties are inevitable and cannot be ignored when modeling a real plant. This has motivated the study of robust performance analysis, robust control and filtering for uncertain singular systems.

In this chapter, we first briefly review results on robustness analysis, robust control and filtering for uncertain singular systems. We then provide a summary of each chapter of this monograph.

1.2 Overview of Robust Control and Filtering of Singular Systems

Roughly speaking, robustness analysis deals with the problem of determining conditions under which the performance or characteristics of a nominal system can be preserved under perturbations of specified types or classes, while robust control/filtering deals with the problem of designing controllers/filters such that the performance or characteristics of the closed-loop/filtering system can be maintained for all allowable parameter uncertainties. One of the most important issues in robustness analysis for singular systems is stability robustness. To illustrate the distinctive features in the stability robustness for singular systems, we consider a nominal singular system described by

$$E\dot{x}(t) = Ax(t), \tag{1.1}$$

where $x(t) \in \mathbb{R}^n$ is the vector of the state variables, and $E, A \in \mathbb{R}^{n \times n}$ are known matrices with E singular. Specifically, we set

$$E = \begin{bmatrix} 1 & 0 & 0 \\ 0 & 0 & 1 \\ 0 & 0 & 0 \end{bmatrix}, \qquad A = \begin{bmatrix} -1 & 0 & 0 \\ 0 & 1 & 0 \\ 0 & 0 & 1 \end{bmatrix}.$$

In this case, it can be verified that (1.1) is regular, stable and has an impulsive mode since $\deg(\det(sE-A)) = 1 \neq \text{rank} E = 2$. Now, we consider the following two perturbation cases:

Case I: Suppose that the derivative matrix E has a perturbation given by the matrix

$$\Delta E = \begin{bmatrix} 0 & 0 & 0 \\ 0 & \epsilon & 0 \\ 0 & 0 & \epsilon \end{bmatrix}.$$

In this case, for any arbitrary small $\epsilon > 0$, it can be verified that $(E + \Delta E, A)$ is unstable since in this case the three finite modes are -1, ϵ^{-1} and ϵ^{-1}. Furthermore, even for the case when (1.1) is regular, impulse-free and stable, it can be shown that one can always choose a perturbation on the derivative matrix E with an arbitrarily small norm to destabilize (1.1).

Case II: Suppose that the matrix A has a perturbation matrix as

$$\Delta A = \begin{bmatrix} 0 & 0 & 0 \\ 0 & 0 & 0 \\ 0 & -\epsilon & 0 \end{bmatrix}.$$

In this case, for any $\epsilon \neq 0$, it can be verified that $(E, A + \Delta A)$ has two finite modes, -1 and ϵ^{-1}, therefore, for any arbitrarily small $\epsilon > 0$, $(E + \Delta E, A)$ is unstable.

These two perturbation cases show that

(I) the stability of the singular system in (1.1) has no tolerance to unstructured uncertainty in the derivative matrix E even if (1.1) is regular, impulse-free and stable. Hence, it is impossible to consider unstructured uncertainties in the derivative matrix E when studying stability robustness;

(II) the stability of the singular system in (1.1) has no tolerance to unstructured uncertainties in the matrix A when the singular system (1.1) is regular and stable but not impulse-free.

We next suppose the singular system (1.1) to be regular, impulse-free and stable, and consider unstructured uncertainty in the matrix A. In this case, the stability as well as the regularity and the non-impulsiveness of (1.1) remain if the norm of the perturbed matrix is within a certain bound. Therefore, stability robustness for singular systems requires considering not only robustness in the finite dynamic modes, but also regularity and non-impulsiveness (for the continuous case) or causality (for the discrete case) at the same time [48, 49]. This is not necessary when considering stability robustness for state-space systems, as the latter two issues do not arise in the state-space case. Therefore, robust stability analysis for singular systems is more complicated than that for state-space systems. Similarly, robust stabilization for uncertain singular systems requires the design of controllers such that the closed-loop system is not only robustly stable but also regular and impulse-free for all allowable parameter uncertainties; this issue is also more complicated than robust stabilization for state-space systems.

In the literature, numerous results on the stability robustness analysis and robust stabilization for singular systems have been reported and various approaches have been proposed. For instance, for singular systems with structured parameter uncertainties, an upper bound of the perturbations was calculated in [48] by using the inverse Laplace transformation and integrals of a modulus matrix function. This upper bound was improved in [105] by applying the maximum modulus principle and spectral radius. By using the technique of Kronecker product and structured singular values, the upper bounds in [48] and [105] were further improved in [111]. The corresponding results for discrete singular systems can be found in [49], where robust stabilizing state feedback controllers were also constructed based on the results on stability robustness. These techniques were adopted to derive robust D-stability conditions in [24, 116, 186]. When both structured parameter uncertainties and time delays appear in a singular system, conditions for stability robustness can be found in [177].

In the case when unstructured parameters are present in a singular system, a suitable measure for the stability radius was defined in [133], where a computable expression for the distance to instability was given. The stability radius for singular systems was also considered in [144], which focused on the derivation of the smallest norm bound on unstructured perturbations to destabilize the system. The exact norm bound for complex perturbations was given, while for real perturbations only several lower bounds were obtained. However, when unidirectional perturbations were considered, exact bounds for such perturbations were derived in [106] by transforming the problem to a rank robustness issue.

In the study of the problems of robust stability analysis and robust stabilization for state-space systems with unstructured parameter uncertainties, it is known that the methods based on the concepts of quadratic stability and quadratic stabilizability have played important roles, for example, [8, 57, 182, 210]. An uncertain state-space system is quadratically stable if there exists a fixed quadratic Lyapunov function which can be used to check the stability of the uncertain system. Accordingly concepts of generalized quadratic stability and generalized quadratic stabilizability have been proposed, to deal respectively with robust stability analysis and robust stabilization problems for uncertain singular systems [179, 190, 192]. Based on these concepts, some robust stability and robust stabilization conditions have been obtained. Similar concepts were also introduced in [113], where perturbation matrices with certain structure in the derivative matrix were considered. These concepts were also extended to uncertain singular delay systems in [188], where necessary and sufficient conditions for generalized quadratic stability and generalized quadratic stabilizability were obtained in terms of linear matrix inequalities (LMIs); these results were also extended to the case with linear fractional parametric uncertainties in [211].

The H_∞ control of state-space systems has been a topic of recurring interest in the past two decades since the H_∞ control method can deal with model uncertainties far more directly (and realistically) than those by classical control and LQ control methods. The purpose of the H_∞ control problem is to design controllers such that the closed-loop system is stable and its H_∞ norm satisfies a prespecified level. A solution to this problem involves solving a set of Riccati equations [41] or LMIs [55]. In the context of singular systems, versions of the bounded real lemma were proposed and H_∞ control results were obtained for both continuous and discrete cases; see, for instance, [80, 125, 157, 170, 191], and the references therein. When parameter uncertainties appear in a singular system, the robust H_∞ control problem was considered in [81], where both state feedback and dynamic output feedback controllers were designed such that the closed-loop system is regular, impulse-free, stable and satisfies a prespecified H_∞ performance level for all allowable parameter uncertainties. Results on robust H_∞ control for uncertain nonlinear singular systems can be found in [69]. When time delays appear in a singular system, the H_∞ control problem was solved in [52, 184, 199], respectively.

Other problems related to robustness analysis and robust control for uncertain singular systems, such as robust controllability and observability, robust normalization, robust output regulation, robust fault-tolerant control, robust positive real control, model predictive control, and guaranteed cost control can be found in [68, 114, 115, 124, 139, 178, 201, 202], respectively.

State estimator design for both state-space and singular systems has been an important research topic in system and control theory. One of the most popular ways to deal with this issue is the celebrated Kalman filtering approach, which generally provides an optimal estimation of the state variables by minimizing the covariance of the estimation error [3]. Under certain assumptions on the structure of singular systems, the Kalman filtering problem was tackled in [35, 38] based on the past dynamics and observations of a given singular system, while in [135] these assumptions were removed and a complete solution to the Kalman filtering problem was obtained by a stochastic shuffle algorithm. The standard Kalman filtering approach usually requires exact information on both the external noises and the internal model of the system. However, these requirements are not always satisfied in practical applications. To overcome these difficulties, an alternative approach called H_∞ filtering has been introduced. In the H_∞ filtering setting, the noise sources are arbitrary signals with bounded energy or average power, and no exact statistics need to be known [130]. It has been shown that the H_∞ filtering technique provides both a guaranteed noise attenuation level and robustness against unmodeled dynamics [13, 130, 147]. The H_∞ filtering problem for state-space systems has been solved via various approaches; see, e.g., [14, 53, 130, 156], and the references therein. In the context of singular systems, [187] seems to be the first paper to investigate the H_∞ filtering problem, where a necessary and sufficient condition for the solvability of this problem was obtained in terms of a set of LMIs. The results in [187] were also extended to the delay case in [198].

1.3 Preview of Chapters

A summary of each chapter of this monograph is given below.

Chapter 2 investigates the problem of stability analysis of singular systems for both continuous and discrete cases. Some basic definitions for singular systems such as regularity, non-impulsiveness (in the continuous case) or causality (in the discrete case) and stability are recalled. A singular system is said to be admissible if it is regular, impulse-free (in the continuous case) or causal (in the discrete case), and stable. Necessary and sufficient conditions for ensuring the admissibility of a singular system are then established via LMIs. These conditions are given in terms of the original system matrices without decomposition.

Chapter 3 focuses on the design of controllers such that the closed-loop system is admissible. In the continuous case, necessary and sufficient conditions for the existence of both stabilizing state feedback and dynamic output feedback controllers are obtained. These conditions are given in terms of LMIs. In the discrete case, a necessary and sufficient condition for the existence of stabilizing state feedback controllers is proposed.

In Chapter 4, the problems of robust stability analysis and robust stabilization for uncertain singular systems are considered. The concepts of generalized quadratic stability and generalized quadratic stabilizability for both continuous- and discrete-time singular systems are proposed. Generalized quadratic stability implies that an uncertain singular system is admissible for all allowable uncertainties. Generalized quadratic stabilizability implies that for an uncertain singular system, there exists a linear state feedback controller such that the closed-loop system is admissible for all allowable uncertainties. Necessary and sufficient conditions for generalized quadratic stability and generalized quadratic stabilizability are derived for both continuous and discrete singular systems with time-invariant norm-bounded parameter uncertainties. Furthermore, for the case with polytopic type uncertainty, sufficient robust stability conditions are proposed in terms of LMIs.

Chapter 5 addresses the H_∞ control problem for singular systems. Versions of the bounded real lemma are proposed for both the continuous and discrete cases. An LMI approach is developed to design state feedback and dynamic output feedback controllers for continuous singular systems such that the closed-loop system is admissible and its transfer function satisfies a prescribed H_∞ performance level. In the context of discrete singular systems, H_∞ state feedback controllers are designed. When parameter uncertainties appear in a singular system model, the problem of robust H_∞ control is investigated. This problem is solved for both continuous and discrete singular systems via the notion of generalized quadratic stabilizability with an H_∞-norm bound.

Chapter 6 deals with the guaranteed cost control problem for singular systems. The purpose is to design a state feedback controller such that the closed-loop system is admissible and a quadratic cost function has an upper bound for all admissible uncertainties. For both continuous and discrete singular systems, sufficient conditions for the solvability of this problem are provided. When these conditions are feasible, desired state feedback controllers can be constructed by solving certain LMIs.

Chapter 7 studies the positive real control problem for singular systems. Versions of the positive real lemma for both continuous and discrete cases are proposed in terms of LMIs. Based on the proposed positive real lemmas, we present necessary and sufficient conditions for the existence of state feedback controllers such that the closed-loop system is admissible and extended strictly positive real in the contexts of both continuous and discrete singular systems.

Chapter 8 investigates the problem of H_∞ filtering for singular systems. First, for continuous singular systems, full-order H_∞ filters are designed such that the resulting error system is admissible and the closed-loop transfer function from the disturbance to the filtering error output satisfies a prescribed H_∞-norm bound constraint. Second, the problem of reduced-order H_∞ filtering is considered for both continuous and discrete singular systems. The purpose is to design a linear proper filter with a specified order lower than the order of the system under consideration, such that the filtering error dynamic system is admissible and its transfer function satisfies a prescribed H_∞ performance level. Necessary and sufficient conditions for the solvability of the reduced-order H_∞ filtering problem are proposed in terms of LMIs and a coupling non-convex rank constraint. As a special case, the problem of the zeroth-order H_∞ filtering can be solved via a convex LMI feasibility problem.

Chapter 9 considers the control and filtering problems for singular systems with time delays. Sufficient conditions for a singular delay system to be regular, impulse-free (for the continuous case) or causal (for the discrete case) and stable are presented in terms of LMIs. Based on these results, stabilizing state feedback controllers are designed for both continuous and discrete singular delay systems. A bounded real lemma is then provided for continuous singular delay systems in terms of LMIs. The H_∞ control and filtering problems are solved by the LMI approach.

Chapter 10 is devoted to the study of Markovian jump singular systems. The concept of stochastic admissibility is introduced for such systems. Necessary and sufficient conditions for stochastic admissibility are obtained in terms of LMIs for continuous and discrete Markovian jump singular systems. The problems of guaranteed cost control and H_∞ control for continuous Markovian jump singular systems are addressed. Sufficient conditions for the solvability of these problems are derived and state feedback controllers are designed via the LMI approach.

2

Stability

2.1 Introduction

This chapter is devoted to the study of the stability for singular systems in both the continuous and discrete contexts. The definitions of regularity and non-impulsiveness (in the continuous case) or causality (in the discrete case) have played important roles in the analysis of singular systems. Specifically, regularity guarantees the existence and uniqueness of a solution to a given singular system, while non-impulsiveness ensures no infinite dynamical modes in a singular system. It is worth mentioning that a singular system usually has three kinds of modes; that is, finite dynamic modes, infinite dynamic modes and infinite nondynamic modes. Infinite dynamic modes can generate undesired impulsive behavior. In the state-space case, neither infinite dynamic modes nor infinite nondynamic can arise. In addition, regularity and non-impulsiveness are satisfied automatically in the state-space systems. Therefore, for a singular system, it is important to develop conditions which guarantee that the given singular system is not only stable but also regular and impulse-free (in the continuous case) or causal (in the discrete case). In the literature, it is reported that such conditions can usually be obtained by decomposing singular systems into slow and fast sub-systems [36]. However, this may lead to some numerical problems. Furthermore, from the mathematical point of view, the decomposition approach is not elegant. Thus, attention in this chapter will be focused on the derivation of such conditions without decomposing the original singular system, and a linear matrix inequality (LMI) approach will be developed.

2.2 Stability of Continuous Systems

Consider a linear continuous singular system described by

$$E\dot{x}(t) = Ax(t) + Bu(t), \tag{2.1}$$
$$y(t) = Cx(t), \tag{2.2}$$

where $x(t) \in \mathbb{R}^n$ is the state; $u(t) \in \mathbb{R}^m$ is the control input; $y(t) \in \mathbb{R}^q$ is the output. The matrix $E \in \mathbb{R}^{n \times n}$ may be singular; we shall assume that $\text{rank}(E) = r \leq n$. A, B and C are known real constant matrices with appropriate dimensions. For the sake of simplicity, sometimes we use (E, A, B, C) to denote the singular system in (2.1) and (2.2).

For the unforced singular system

$$E\dot{x}(t) = Ax(t), \tag{2.3}$$

or the pair (E, A), we define its generalized spectral abscissa as

$$\alpha(E, A) \triangleq \max_{\lambda \in \{s | \det(sE - A) = 0\}} \text{Re}(\lambda).$$

For notational simplicity, we also write $\alpha(A) = \alpha(I, A)$ which is the usual spectral abscissa. We introduce the following definition.

Definition 2.1. [36, 108]

(I) *The pair (E, A) is said to be regular if $\det(sE - A)$ is not identically zero.*

(II) *The pair (E, A) is said to be impulse-free if $\deg(\det(sE - A)) = \text{rank}(E)$.*

(III) *The pair (E, A) is said to be stable if all the roots of $\det(sE - A) = 0$ have negative real parts.*

(IV) *The pair (E, A) is said to be admissible if it is regular, impulse-free and stable.*

Remark 2.1. It is noted that the regularity of the pair (E, A) guarantees the existence and uniqueness of a solution to (2.3) for any specified initial conditions. Furthermore, from Definition 2.1, it can be deduced that the non-impulsiveness of the pair (E, A) implies that the pair (E, A) is regular. ◁

The following lemma gives an equivalent condition for regularity.

Lemma 2.1 (Slow-Fast Decomposition). [36] *The pair (E, A) is regular if and only if there exist two nonsingular matrices M_1 and N_1 such that*

$$M_1 E N_1 = \text{diag}(I, \mathcal{J}), \quad M_1 A N_1 = \text{diag}(\mathcal{A}, I), \qquad (2.4)$$

where $\mathcal{J}$ is a nilpotent matrix.

Lemma 2.2. [36] *Suppose that the pair (E, A) is regular, and two nonsingular matrices M_1 and N_1 are found such that (2.4) holds, then we have:*

(a) The pair (E, A) is impulse-free if and only if $\mathcal{J} = 0$.

(b) The pair (E, A) is stable if and only if $\alpha(\mathcal{A}) < 0$.

(c) The pair (E, A) is admissible if and only if $\mathcal{J} = 0$ and $\alpha(\mathcal{A}) < 0$.

When the regularity of the pair (E, A) is not known, it is always possible to choose two nonsingular matrices M_2 and N_2 such that

$$M_2 E N_2 = \begin{bmatrix} I & 0 \\ 0 & 0 \end{bmatrix}, \quad M_2 A N_2 = \begin{bmatrix} A_1 & A_2 \\ A_3 & A_4 \end{bmatrix}. \qquad (2.5)$$

The decomposition can be obtained via a singular value decomposition on E and followed by scaling of the bases. Then, we have the following result.

Lemma 2.3. [36]

(a) The pair (E, A) is impulse-free if and only if A_4 is nonsingular.

(b) The pair (E, A) is admissible if and only if A_4 is nonsingular and

$$\alpha \left(A_1 - A_2 A_4^{-1} A_3 \right) < 0.$$

Remark 2.2. Both Lemmas 2.2 and 2.3 present equivalent conditions on the admissibility of the pair (E, A). Lemma 2.2 is based on the assumption that the pair (E, A) is regular. It is noted that both conditions involve the decomposition of the matrices of the original singular system; that is, we have to find two nonsingular matrices M_1 and N_1, and M_2 and N_2 such that (2.4) and (2.5) are satisfied, respectively, which is sometimes numerically unreliable, especially in the case when the order of the system is relatively large.

◁

In view of Remark 2.2, next we will present another equivalent admissibility condition in terms of LMIs, which involves no decomposition of the system matrices. Before giving such a condition, we first introduce the following results, which will be used in the sequel.

Lemma 2.4. [40] *Given any real square matrix $\mathcal{X}$ with appropriate dimensions. The matrix measure $\mu(\mathcal{X})$ defined as*

$$\mu(\mathcal{X}) = \lim_{\theta \to 0^+} \frac{\|I + \theta\mathcal{X}\| - 1}{\theta}$$

has the following properties:

(I)

$$-\|\mathcal{X}\| \leq \alpha(\mathcal{X}) \leq \mu(\mathcal{X}) \leq \|\mathcal{X}\|.$$

(II)

$$\mu(\mathcal{X}) = \frac{1}{2}\lambda_{\max}\left(\mathcal{X} + \mathcal{X}^T\right) = \frac{1}{2}\alpha\left(\mathcal{X} + \mathcal{X}^T\right).$$

Lemma 2.5 (Matrix Inversion Lemma). [101] *Given any matrices A, B, C and D with appropriate dimensions such that A, C and $A + BCD$ are nonsingular. Then we have*

$$(A + BCD)^{-1} = A^{-1} - A^{-1}B\left(DA^{-1}B + C^{-1}\right)^{-1}DA^{-1}.$$

Lemma 2.6 (Schur Complement). [15] *Given any real matrices $\mathcal{P}_1$, $\mathcal{P}_2$ and $\mathcal{P}_3$ with $\mathcal{P}_1 = \mathcal{P}_1^T$ and $\mathcal{P}_3 > 0$. Then we have*

$$\mathcal{P}_1 + \mathcal{P}_2\mathcal{P}_3^{-1}\mathcal{P}_2^T < 0,$$

if and only if

$$\begin{bmatrix} \mathcal{P}_1 & \mathcal{P}_2 \\ \mathcal{P}_2^T & -\mathcal{P}_3 \end{bmatrix} < 0,$$

or equivalently

$$\begin{bmatrix} -\mathcal{P}_3 & \mathcal{P}_2^T \\ \mathcal{P}_2 & \mathcal{P}_1 \end{bmatrix} < 0.$$

Lemma 2.7. [88] *Let $\mathcal{X}$ and $\mathcal{Y}$ be any given real matrices of appropriate dimensions. Then, for any scalar $\epsilon > 0$,*

$$\mathcal{X}^T\mathcal{Y} + \mathcal{Y}^T\mathcal{X} \leq \epsilon^{-1}\mathcal{X}^T\mathcal{X} + \epsilon\mathcal{Y}^T\mathcal{Y}.$$

Lemma 2.8. *Let*

$$\mathcal{N} = \begin{bmatrix} \mathcal{P} & \mathcal{X} \\ \mathcal{Y} & \mathcal{Z} \end{bmatrix},$$

where $\mathcal{P}$, $\mathcal{X}$, $\mathcal{Y}$ and $\mathcal{Z}$ are any real given matrices with appropriate dimensions such that

$$\mathcal{N} + \mathcal{N}^T < 0. \tag{2.6}$$

Then, $\mathcal{Z}$ is nonsingular and

$$\mathcal{P} + \mathcal{P}^T - \mathcal{X} \mathcal{Z}^{-1} \mathcal{Y} - \mathcal{Y}^T \mathcal{Z}^{-T} \mathcal{X}^T < 0. \tag{2.7}$$

Proof. By (2.6), it is easy to see that

$$\mathcal{Z} + \mathcal{Z}^T < 0. \tag{2.8}$$

Then, applying Lemma 2.4, we have $\mu(\mathcal{Z}) < 0$, which implies $\alpha(\mathcal{Z}) < 0$. Therefore $\mathcal{Z}$ is nonsingular. Pre- and post-multiplying (2.8) by $\mathcal{Z}^{-1}$ and $\mathcal{Z}^{-T}$, respectively, result in

$$\mathcal{Z}^{-1} + \mathcal{Z}^{-T} < 0, \tag{2.9}$$

so, $\mathcal{Z}^{-1} + \mathcal{Z}^{-T}$ is invertible. Set

$$H = -\left(\mathcal{Z}^{-1} + \mathcal{Z}^{-T} \right)^{-1}.$$

Then, from (2.9), it is easy to see that $H > 0$, which implies that there exists a nonsingular matrix W such that

$$H = W^T W.$$

By Lemma 2.7, it can be deduced that

$$
\begin{aligned}
\mathcal{X} \mathcal{Z}^{-1} &H \mathcal{Z}^{-1} \mathcal{Y} + \mathcal{Y}^T \mathcal{Z}^{-T} H^T \mathcal{Z}^{-T} \mathcal{X}^T \\
&= \left(\mathcal{X} \mathcal{Z}^{-1} W^T \right) \left(W \mathcal{Z}^{-1} \mathcal{Y} \right) + \left(\mathcal{Y}^T \mathcal{Z}^{-T} W^T \right) \left(W \mathcal{Z}^{-T} \mathcal{X}^T \right) \\
&\leq \left(\mathcal{X} \mathcal{Z}^{-1} W^T \right) \left(W \mathcal{Z}^{-T} \mathcal{X}^T \right) + \left(\mathcal{Y}^T \mathcal{Z}^{-T} W^T \right) \left(W \mathcal{Z}^{-1} \mathcal{Y} \right) \\
&= -\mathcal{X} \left(\mathcal{Z} + \mathcal{Z}^T \right)^{-1} \mathcal{X}^T - \mathcal{Y}^T \left(\mathcal{Z} + \mathcal{Z}^T \right)^{-1} \mathcal{Y}. \tag{2.10}
\end{aligned}
$$

That is,

$$
\begin{aligned}
-\mathcal{X} \left(\mathcal{Z} + \mathcal{Z}^T \right)^{-1} &\mathcal{X}^T - \mathcal{Y}^T \left(\mathcal{Z} + \mathcal{Z}^T \right)^{-1} \mathcal{Y} \\
&-\mathcal{X} \mathcal{Z}^{-1} H \mathcal{Z}^{-1} \mathcal{Y} - \mathcal{Y}^T \mathcal{Z}^{-T} H^T \mathcal{Z}^{-T} \mathcal{X}^T \geq 0. \tag{2.11}
\end{aligned}
$$

Furthermore, using Lemma 2.5, we obtain

$$
\begin{aligned}
\mathcal{X} \mathcal{Z}^{-1} \mathcal{Y} &+ \mathcal{Y}^T \mathcal{Z}^{-T} \mathcal{X}^T - \left(\mathcal{X} + \mathcal{Y}^T \right) \left(\mathcal{Z} + \mathcal{Z}^T \right)^{-1} \left(\mathcal{X} + \mathcal{Y}^T \right)^T \\
&= -\mathcal{X} \left(\mathcal{Z} + \mathcal{Z}^T \right)^{-1} \mathcal{X}^T - \mathcal{Y}^T \left(\mathcal{Z} + \mathcal{Z}^T \right)^{-1} \mathcal{Y} \\
&-\mathcal{X} \mathcal{Z}^{-1} H \mathcal{Z}^{-1} \mathcal{Y} - \mathcal{Y}^T \mathcal{Z}^{-T} H^T \mathcal{Z}^{-T} \mathcal{X}^T.
\end{aligned}
$$

This together with (2.11) implies

$$\mathcal{X}\mathcal{Z}^{-1}\mathcal{Y} + \mathcal{Y}^T \mathcal{Z}^{-T}\mathcal{X}^T \geq \left(\mathcal{X} + \mathcal{Y}^T\right)\left(\mathcal{Z} + \mathcal{Z}^T\right)^{-1}\left(\mathcal{X} + \mathcal{Y}^T\right)^T. \tag{2.12}$$

On the other hand, noting (2.6) and using Lemma 2.6, we obtain

$$\mathcal{P} + \mathcal{P}^T - \left(\mathcal{X} + \mathcal{Y}^T\right)\left(\mathcal{Z} + \mathcal{Z}^T\right)^{-1}\left(\mathcal{X} + \mathcal{Y}^T\right)^T < 0. \tag{2.13}$$

Finally, it follows from (2.12) and (2.13) that (2.7) is satisfied. This completes the proof.   □

Now, we are in a position to present a necessary and sufficient condition for the singular system in (2.3) to be admissible via LMIs which has also been reported in [125]. Here, we offer a different derivation.

Theorem 2.1. *The pair (E, A) is admissible if and only if there exists a matrix P such that*

$$E^T P = P^T E \geq 0, \tag{2.14}$$
$$P^T A + A^T P < 0. \tag{2.15}$$

Proof. (Sufficiency) Assume that there exists a matrix P such that (2.14) and (2.15) are satisfied. Under this condition, we first show that the pair (E, A) is regular and impulse-free. To this end, we choose two nonsingular matrices M and N such that

$$E = M \begin{bmatrix} I & 0 \\ 0 & 0 \end{bmatrix} N, \quad A = M \begin{bmatrix} A_1 & A_2 \\ A_3 & A_4 \end{bmatrix} N. \tag{2.16}$$

Write

$$M^T P N^{-1} = \begin{bmatrix} P_1 & P_2 \\ P_3 & P_4 \end{bmatrix}, \tag{2.17}$$

where the partition is compatible with that of A in (2.16). Then, by (2.14), it can be shown that $P_2 = 0$, and

$$P_1 = P_1^T \geq 0. \tag{2.18}$$

Noting (2.15) and using Lemma 2.4, we have

$$\alpha(A^T P) \leq \mu\left(A^T P\right) = \frac{1}{2}\lambda_{\max}\left(P^T A + A^T P\right) < 0.$$

Hence, $A^T P$ is nonsingular, which implies that P is nonsingular too. This together with (2.18) results in $P_1 > 0$. Now, pre- and post-multiplying (2.15) by N^{-T} and N^{-1}, respectively, and then using the expressions in (2.16) and (2.17), we have

$$\begin{bmatrix} U_1 & U_2 \\ U_2^T & U_3 \end{bmatrix} < 0, \tag{2.19}$$

where

$$\begin{aligned}
U_1 &= A_1^T P_1 + P_1^T A_1 + A_3^T P_3 + P_3^T A_3, \\
U_2 &= A_3^T P_4 + P_1^T A_2 + P_3^T A_4, \\
U_3 &= A_4^T P_4 + P_4^T A_4.
\end{aligned}$$

Then, the 2-2 block in (2.19) gives

$$A_4^T P_4 + P_4^T A_4 < 0.$$

This, by Lemma 2.4, gives

$$\alpha(P_4^T A_4) \le \mu\left(P_4^T A_4\right) = \frac{1}{2}\lambda_{\max}\left(A_4^T P_4 + P_4^T A_4\right) < 0,$$

so, $P_4^T A_4$ is nonsingular, which implies A_4 is nonsingular too. Thus, by Lemmas 2.2 and 2.3, it can be seen that the pair (E, A) is regular and impulse-free. Next, we show the stability of the pair (E, A). To this end, we define

$$\hat{N} = \begin{bmatrix} P_1^T A_1 + P_3^T A_3 & A_3^T P_4 \\ A_2^T P_1 + A_4^T P_3 & A_4^T P_4 \end{bmatrix}.$$

Considering (2.19), it is easy to see that

$$\hat{N} + \hat{N}^T = \begin{bmatrix} U_1 & U_2 \\ U_2^T & U_3 \end{bmatrix} < 0, \tag{2.20}$$

Therefore, by Lemma 2.8, it follows that

$$\left(A_1 - A_2 A_4^{-1} A_3\right)^T P_1 + P_1 \left(A_1 - A_2 A_4^{-1} A_3\right) < 0.$$

Noting $P_1 > 0$ and employing the Lyapunov stability theory in [4], we have

$$\alpha\left(A_1 - A_2 A_4^{-1} A_3\right) < 0.$$

Then, applying Lemma 2.3, we have that the pair (E, A) is stable. This together with the regularity and non-impulsiveness of the pair (E, A) gives that the pair (E, A) is admissible.

(*Necessity*) Assume that the pair (E, A) is admissible. Then, by Lemmas 2.1 and 2.2, we have that there exist two nonsingular matrices M_1 and N_1 satisfying (2.4) with $\mathcal{J} = 0$; that is,

$$M_1 E N_1 = \begin{bmatrix} I & 0 \\ 0 & 0 \end{bmatrix}, \quad M_1 A N_1 = \begin{bmatrix} \mathcal{A} & 0 \\ 0 & I \end{bmatrix}, \tag{2.21}$$

in which $\alpha(\mathcal{A}) < 0$. Then, by the Lyapunov stability theory in [4], it can be seen that there exists a matrix $\mathcal{P}_1 > 0$ such that

$$\mathcal{A}^T \mathcal{P}_1 + \mathcal{P}_1 \mathcal{A} < 0. \tag{2.22}$$

Set

$$P = M_1^T \begin{bmatrix} \mathcal{P}_1 & 0 \\ 0 & -I \end{bmatrix} N_1^{-1}. \tag{2.23}$$

Then, considering (2.21) and (2.22), it is easy to see that the matrix P defined in (2.23) satisfies (2.14) and (2.15). This completes the proof. $\quad\square$

By Theorem 2.1, we obtain the following corollary.

Corollary 2.1. *The pair (E, A) is admissible if and only if there exists a matrix P such that*

$$EP = P^T E^T \geq 0, \tag{2.24}$$
$$P^T A^T + AP < 0. \tag{2.25}$$

Proof. By Lemmas 2.1 and 2.2, it is to see that the pair (E, A) is admissible if and only if the pair (E^T, A^T) is admissible. Then the desired result follows immediately by using Theorem 2.1. $\quad\square$

Remark 2.3. Both Theorem 2.1 and Corollary 2.1 give necessary and sufficient conditions for the pair (E, A) to be admissible in terms of LMIs. However, it is noted that the conditions in (2.14) and (2.24) are non-strict LMIs, which contain equality constraints; this may result in numerical problems when checking such non-strict LMI conditions since equality constraints are fragile and usually not satisfied perfectly. Therefore, strict LMI conditions are more desirable than non-strict ones from the numerical point of view. $\quad\triangleleft$

Considering Remark 2.3, we give a strict LMI condition for admissibility in the following theorem.

Theorem 2.2. *The pair (E, A) is admissible if and only if there exist matrices $P > 0$ and Q such that*

$$(PE + SQ)^T A + A^T (PE + SQ) < 0, \tag{2.26}$$

where $S \in \mathbb{R}^{n \times (n-r)}$ is any matrix with full column rank and satisfies $E^T S = 0$.

Proof. (*Sufficiency*) Assume that there exist matrices $P > 0$ and Q such that (2.26) holds. Let

$$\mathcal{P} = PE + SQ.$$

Then, by (2.26), it is easy to see

$$E^T \mathcal{P} = \mathcal{P}^T E = E^T PE \geq 0, \tag{2.27}$$

$$\mathcal{P}^T A + A^T \mathcal{P} < 0. \tag{2.28}$$

Therefore, by Theorem 2.1, we have that the pair (E, A) is admissible.

(*Necessity*) Assume that the pair (E, A) is admissible. Then, by Lemmas 2.1 and 2.2, we have that there exist two nonsingular matrices $\mathcal{M}_1$ and $\mathcal{N}_1$ such that

$$\mathcal{M}_1 E \mathcal{N}_1 = \begin{bmatrix} I & 0 \\ 0 & 0 \end{bmatrix}, \quad \mathcal{M}_1 A \mathcal{N}_1 = \begin{bmatrix} \mathcal{A} & 0 \\ 0 & I \end{bmatrix}, \tag{2.29}$$

where

$$\alpha(\mathcal{A}) < 0. \tag{2.30}$$

Considering the definition of the matrix S, we can write

$$S = \mathcal{M}_1^T \begin{bmatrix} 0 \\ I \end{bmatrix} H,$$

where H is any nonsingular matrix. Noting (2.30) and applying the Lyapunov stability theory in [4], it can be seen that there exists a matrix $\mathcal{P}_1 > 0$ such that

$$\mathcal{A}^T \mathcal{P}_1 + \mathcal{P}_1 \mathcal{A} < 0. \tag{2.31}$$

Set

$$P = \mathcal{M}_1^T \begin{bmatrix} \mathcal{P}_1 & 0 \\ 0 & I \end{bmatrix} \mathcal{M}_1, \quad Q = H^{-1} \begin{bmatrix} 0 & -I \end{bmatrix} \mathcal{N}_1^{-1}. \tag{2.32}$$

Then, it can be verified that the matrices P and Q defined in (2.32) satisfy (2.26). This completes the proof. □

Remark 2.4. In the case when $E = I$, that is, the unforced singular system in (2.3) reduces to a state-space system, it is easy to see that $S = 0$; then, Theorem 2.2 coincides with the Lyapunov stability theory in [4]. Therefore, Theorem 2.2 can be regarded as an extension of the Lyapunov stability theory for continuous state-space systems to continuous singular systems. ◁

By Theorem 2.2, it is easy to obtain the following corollary.

Corollary 2.2. *The pair* (E, A) *is admissible if and only if there exist matrices* $P > 0$ *and* Q *such that*

$$\left(PE^T + SQ\right)^T A^T + A\left(PE^T + SQ\right) < 0, \tag{2.33}$$

where $S \in \mathbb{R}^{n \times (n-r)}$ is any matrix with full column rank and satisfies $ES = 0$.

To illustrate the results in Theorem 2.2, we now provide a numerical example.

Example 2.1. Consider an unforced continuous singular system in (2.3) with

$$E = \begin{bmatrix} 1 & 1 & 0.5 \\ -0.5 & 1.5 & 1.75 \\ 1 & 1 & 0.5 \end{bmatrix}, \quad A = \begin{bmatrix} -10 & 5 & 6.5 \\ 2 & -5.5 & -1.25 \\ -9 & 4 & 8.5 \end{bmatrix}.$$

We choose

$$S = \begin{bmatrix} 1 \\ 0 \\ -1 \end{bmatrix},$$

which is with full column rank and satisfies $E^T S = 0$. Then, by solving the LMI in (2.26), we obtain the solution as follows:

$$P = \begin{bmatrix} 0.5249 & 0.0070 & -0.4954 \\ 0.0070 & 0.0996 & 0.0070 \\ -0.4954 & 0.0070 & 0.5249 \end{bmatrix},$$

$$Q = \begin{bmatrix} -0.0334 & -0.0003 & 0.3464 \end{bmatrix}.$$

Therefore, by Theorem 2.2, it is easy to see that the singular system in this example is admissible.

The admissibility of this singular system can also be verified by using Lemma 2.2. To this end, we choose two nonsingular matrices

$$M = \begin{bmatrix} 2 & -1 & 1 \\ 1 & 0.5 & -1.5 \\ 2 & -1 & 0 \end{bmatrix}, \quad N = \begin{bmatrix} 0 & 1 & 1 \\ -1 & 1 & 1.5 \\ -1 & 1 & -2 \end{bmatrix},$$

such that

$$E = M \begin{bmatrix} I_2 & 0 \\ 0 & 0 \end{bmatrix} N, \quad A = M \begin{bmatrix} \mathcal{A} & 0 \\ 0 & 1 \end{bmatrix} N,$$

where

$$\mathcal{A} = \begin{bmatrix} -3 & 2 \\ -1 & -5 \end{bmatrix}.$$

Then, it can be shown that the two eigenvalues of the matrix $\mathcal{A}$ are $-4 + j$ and $-4 - j$; that is, $\alpha(\mathcal{A}) < 0$. Hence, by Lemma 2.2, it is easy to conclude that the singular system is admissible.   ◇

2.3 Stability of Discrete Systems

Consider a linear discrete singular system described by

$$Ex(k+1) = Ax(k) + Bu(k), \qquad (2.34)$$
$$y(k) = Cx(k), \qquad (2.35)$$

where $x(k) \in \mathbb{R}^n$ is the state; $u(k) \in \mathbb{R}^m$ is the control input; $y(k) \in \mathbb{R}^q$ is the output. The matrix $E \in \mathbb{R}^{n \times n}$ may be singular; we shall assume that rank$(E) = r \leq n$. A, B and C are known real constant matrices with appropriate dimensions. Similar to the continuous case, we use (E, A, B, C) to denote the singular system in (2.34) and (2.35).

For the unforced singular system

$$Ex(k+1) = Ax(k), \qquad (2.36)$$

or the pair (E, A), we define its generalized spectral radius as

$$\rho(E, A) \triangleq \max_{\lambda \in \{z \mid \det(zE - A) = 0\}} |\lambda|$$

For notational simplicity, we also write $\rho(A) = \rho(I, A)$ which is the usual spectral radius. We introduce the following definition.

Definition 2.2. [36, 108]

(I) The pair (E, A) is said to be regular if $\det(zE - A)$ is not identically zero.

(II) The pair (E, A) is said to be causal if $\deg(\det(zE - A)) = \mathrm{rank}(E)$.

(III) The pair (E, A) is said to be stable if $\rho(E, A) < 1$.

(IV) The pair (E, A) is said to be admissible if it is regular, causal and stable.

Suppose that the pair (E, A) is regular, then there exist two nonsingular matrices M_1 and N_1 such that (2.4) holds. In this case, we have the following result.

Lemma 2.9.

(a) The pair (E, A) is causal if and only if $\mathcal{J} = 0$.

(b) The pair (E, A) is stable if and only if $\rho(\mathcal{A}) < 1$.

(c) The pair (E, A) is admissible if and only if $\mathcal{J} = 0$ and $\rho(\mathcal{A}) < 1$.

For any given pair (E, A), there always exist two invertible matrices M_2 and N_2 such that (2.5) holds. Then, we have the following result.

Lemma 2.10.

(a) The pair (E, A) is causal if and only if A_4 is nonsingular.

(b) The pair (E, A) is admissible if and only if A_4 is nonsingular and

$$\rho(A_1 - A_2 A_4^{-1} A_3) < 1.$$

Both Lemmas 2.9 and 2.10 give equivalent conditions on the admissibility of discrete singular systems. However, these conditions are based on the decomposition of system matrices, which may result in numerical problems and lacks theoretical elegance. Considering this, we present another admissibility condition in the following theorem which involves no decomposition of system matrices.

Theorem 2.3. *The pair (E, A) is admissible if and only if there exists a matrix $P = P^T$ such that*

$$E^T P E \geq 0, \tag{2.37}$$
$$A^T P A - E^T P E < 0. \tag{2.38}$$

In this case, the matrix P is nonsingular.

Proof. (*Necessity*) Suppose that the pair (E, A) is admissible, then it follows from Lemma 2.9 that there exist two nonsingular matrices M and N such that

$$E = M \begin{bmatrix} I & 0 \\ 0 & 0 \end{bmatrix} N, \quad A = M \begin{bmatrix} \bar{A} & 0 \\ 0 & I \end{bmatrix} N. \tag{2.39}$$

where $\bar{A} \in \mathbb{R}^{r \times r}$ and

$$\rho(\bar{A}) < 1. \tag{2.40}$$

By using Lyapunov stability theory, it follows that there exists a matrix $\bar{P} > 0$ such that

$$\bar{A}^T \bar{P} \bar{A} - \bar{P} < 0. \tag{2.41}$$

Now, define a matrix P by

$$P = M^{-T} \begin{bmatrix} \bar{P} & 0 \\ 0 & -I \end{bmatrix} M^{-1}. \tag{2.42}$$

Then, it is easy to verify that the symmetric matrix given in (2.42) satisfies (2.38).

(*Sufficiency*) Let $\hat{M}$ and $\hat{N}$ be two nonsingular matrices such that

$$E = \hat{M} \begin{bmatrix} I & 0 \\ 0 & 0 \end{bmatrix} \hat{N}. \tag{2.43}$$

Write

$$\hat{M}^T P \hat{M} = \begin{bmatrix} \hat{P}_1 & \hat{P}_2 \\ \hat{P}_2^T & \hat{P}_3 \end{bmatrix}, \quad A = \hat{M} \begin{bmatrix} A_1 & A_2 \\ A_3 & A_4 \end{bmatrix} \hat{N}, \tag{2.44}$$

where the partition is compatible with that of E in (2.44). From (2.37), (2.43) and (2.44), it is easy to show that

$$\hat{P}_1 \geq 0. \tag{2.45}$$

Substituting (2.43) and (2.44) into (2.38) gives

$$\hat{N}^T \begin{bmatrix} * & * \\ * & A_2^T \hat{P}_1 A_2 + \mathcal{H} + \mathcal{H}^T \end{bmatrix} \hat{N} < 0, \tag{2.46}$$

where $*$ represents matrices that are not relevant in the following discussion, and

$$\mathcal{H} = A_2^T \hat{P}_2 A_4 + \frac{1}{2} A_4^T \hat{P}_3 A_4.$$

Taking into account (2.46) and noting (2.45), we obtain

$$\mathcal{H} + \mathcal{H}^T < 0. \tag{2.47}$$

Then, by Lemma 2.4, it follows that

$$\alpha \left(\left(A_2^T \hat{P}_2 + \frac{1}{2} A_4^T \hat{P}_3 \right) A_4 \right)$$
$$= \alpha(\mathcal{H}) \leq \mu(\mathcal{H})$$
$$= \frac{1}{2} \lambda_{\max} \left(\mathcal{H} + \mathcal{H}^T \right) < 0.$$

Then, it can be easily shown that the matrix $\left(A_2^T \hat{P}_2 + \frac{1}{2} A_4^T \hat{P}_3 \right) A_4$ is nonsingular, which implies that the matrix A_4 is nonsingular too. Then, by Lemmas 2.9 and 2.10, we have that the pair (E, A) is regular and causal.

On the other hand, the regularity and causality of the pair (E, A) imply that there exist two nonsingular matrices $\tilde{M}$ and $\tilde{N}$ such that

$$E = \tilde{M} \begin{bmatrix} I & 0 \\ 0 & 0 \end{bmatrix} \tilde{N}, \quad A = \tilde{M} \begin{bmatrix} \tilde{A} & 0 \\ 0 & I \end{bmatrix} \tilde{N}. \tag{2.48}$$

Write

$$\tilde{M}^T P \tilde{M} = \begin{bmatrix} \tilde{P}_1 & \tilde{P}_2 \\ \tilde{P}_2^T & \tilde{P}_3 \end{bmatrix}, \tag{2.49}$$

where the partition is compatible with that of A in (2.48). From (2.37) and (2.49), it is easy to show that

$$\tilde{P}_1 \geq 0. \tag{2.50}$$

Then, using (2.38) together with (2.48) and (2.49) result in

$$\tilde{A}\tilde{P}_1\tilde{A}^T - \tilde{P}_1 < 0. \tag{2.51}$$

By (2.50) and (2.51), it can be deduced that

$$\tilde{P}_1 > 0. \tag{2.52}$$

This together with (2.51) implies

$$\rho(\tilde{A}) < 1 \tag{2.53}$$

Therefore, the pair (E, A) is stable. This together with the regularity and causality of the pair (E, A) gives that the pair (E, A) is admissible.

Next, we show that P is nonsingular when (2.37) and (2.38) hold. To this end, we substitute (2.48) and (2.49) into (2.38) and obtain

$$\tilde{N}^T \begin{bmatrix} \tilde{A}^T \tilde{P}_1 \tilde{A} - \tilde{P}_1 & \tilde{A}^T \tilde{P}_2 \\ \tilde{P}_2^T \tilde{A} & \tilde{P}_3 \end{bmatrix} \tilde{N} < 0,$$

which, by Schur complement, yields

$$\tilde{P}_3 < 0, \tag{2.54}$$

and

$$\tilde{A}^T \left(\tilde{P}_1 - \tilde{P}_2 \tilde{P}_3^{-1} \tilde{P}_2^T \right) \tilde{A} - \tilde{P}_1 < 0. \tag{2.55}$$

Let

$$\tilde{W} = \tilde{P}_1 - \tilde{P}_2 \tilde{P}_3^{-1} \tilde{P}_2^T. \tag{2.56}$$

Then, from (2.54) and (2.55), it can be seen that

$$\tilde{A}^T \tilde{W} \tilde{A} - \tilde{W} < 0. \tag{2.57}$$

That is, there exists a matrix $\tilde{Q} > 0$ such that

$$\tilde{A}^T \tilde{W} \tilde{A} - \tilde{W} + \tilde{Q} = 0. \tag{2.58}$$

This together with (2.53) gives

$$\tilde{W} = \sum_{i=1}^{\infty} \left(\tilde{A}^T\right)^i \tilde{Q}\tilde{A} + \tilde{Q} > 0.$$

Therefore, $\tilde{P}_1 - \tilde{P}_2\tilde{P}_3^{-1}\tilde{P}_2^T$ is nonsingular. Recalling $\tilde{P}_3$ is nonsingular and noting (2.49), we have that $\tilde{M}^T P \tilde{M}$ is nonsingular, and thus P is nonsingular too. This completes the proof. $\square$

The following theorem provides a strict LMI condition on the admissibility of discrete singular systems.

Theorem 2.4. *The pair (E, A) is admissible if and only if there exist matrices $P > 0$ and Q such that*

$$A^T P A - E^T P E + Q S^T A + A^T S Q^T < 0, \qquad (2.59)$$

where $S \in \mathbb{R}^{n \times (n-r)}$ is any matrix with full column rank and satisfies $E^T S = 0$.

Proof. (*Necessity*) Suppose that the pair (E, A) is admissible, then it follows from Lemmas 2.9 and 2.10 that there exist two nonsingular matrices M and N such that

$$E = M \begin{bmatrix} I & 0 \\ 0 & 0 \end{bmatrix} N, \quad A = M \begin{bmatrix} \bar{A} & 0 \\ 0 & I \end{bmatrix} N. \qquad (2.60)$$

where $\bar{A} \in \mathbb{R}^{r \times r}$ and

$$\rho(\bar{A}) < 1. \qquad (2.61)$$

Then, S can be expressed as

$$S = M^{-T} \begin{bmatrix} 0 \\ I \end{bmatrix} H, \qquad (2.62)$$

where H is any nonsingular matrix. Considering (2.61), we have that there exists a matrix $\bar{P} > 0$ such that

$$\bar{A}^T \bar{P} \bar{A} - \bar{P} < 0.$$

Now, define

$$P = M^{-T} \begin{bmatrix} \bar{P} & 0 \\ 0 & I \end{bmatrix} M^{-1}, \quad Q = N^T \begin{bmatrix} 0 \\ -I \end{bmatrix} H^{-T}. \qquad (2.63)$$

Then, it is easy to verify that the matrices P and Q in (2.63) satisfy (2.59).

(*Sufficiency*) Assume that there exist matrices $P > 0$ and Q such that (2.59) is satisfied. We first show that the pair (E, A) is regular and causal. To this end, we choose two nonsingular matrices $\hat{M}$ and $\hat{N}$ such that

$$E = \hat{M} \begin{bmatrix} I & 0 \\ 0 & 0 \end{bmatrix} \hat{N}, \quad A = \hat{M} \begin{bmatrix} A_1 & A_2 \\ A_3 & A_4 \end{bmatrix} \hat{N}. \tag{2.64}$$

Then S can be given as

$$S = \hat{M}^{-T} \begin{bmatrix} 0 \\ I \end{bmatrix} \hat{H}, \tag{2.65}$$

where $\hat{H}$ is any nonsingular matrix. Write

$$\hat{M}^T P \hat{M} = \begin{bmatrix} \hat{P}_1 & \hat{P}_2 \\ \hat{P}_2^T & \hat{P}_3 \end{bmatrix}, \quad \hat{N}^{-T} Q \hat{H}^T = \begin{bmatrix} Q_1 \\ Q_2 \end{bmatrix}, \tag{2.66}$$

where the partition is compatible with that of A in (2.64). Now, substituting (2.64)–(2.66) to (2.59) gives

$$\hat{M} \begin{bmatrix} * & * \\ * & W \end{bmatrix} \hat{M}^T < 0, \tag{2.67}$$

where $*$ represents matrices that are not relevant in the following discussion, and

$$W = A_2^T \hat{P}_1 A_2 + A_4^T \hat{P}_2^T A_2 + A_2^T \hat{P}_2 A_4 + A_4^T \hat{P}_3 A_4 + Q_2 A_4 + A_4^T Q_2^T.$$

From (2.67), it is easy to see

$$W < 0. \tag{2.68}$$

That is,

$$\left(A_2^T \hat{P}_2 + \frac{1}{2} A_4^T \hat{P}_3 + Q_2 \right) A_4 + A_4^T \left(A_2^T \hat{P}_2 + \frac{1}{2} A_4^T \hat{P}_3 + Q_2 \right)^T < 0,$$

which, by Lemma 2.4, gives

$$\alpha \left(\left(A_2^T \hat{P}_2 + \frac{1}{2} A_4^T \hat{P}_3 + Q_2 \right) A_4 \right)$$

$$\leq \mu \left(\left(A_2^T \hat{P}_2 + \frac{1}{2} A_4^T \hat{P}_3 + Q_2 \right) A_4 \right)$$

$$= \frac{1}{2} \lambda_{\max} \left(\left(A_2^T \hat{P}_2 + \frac{1}{2} A_4^T \hat{P}_3 + Q_2 \right) A_4 \right.$$

$$\left. + A_4^T \left(A_2^T \hat{P}_2 + \frac{1}{2} A_4^T \hat{P}_3 + Q_2 \right)^T \right)$$

$$< 0.$$

Therefore, the matrix $\left(A_2^T \hat{P}_2 + \frac{1}{2} A_4^T \hat{P}_3 + Q_2\right) A_4$ is nonsingular, which implies that the matrix A_4 is nonsingular too. Then, by Lemmas 2.9 and 2.10, we have that the pair (E, A) is regular and causal.

Next, we show that the pair (E, A) is stable. Considering that the pair (E, A) is regular and causal, by Lemmas 2.9 and 2.10, it is easy to see that there exist two nonsingular matrices $\tilde{M}$ and $\tilde{N}$ such that

$$E = \tilde{M} \begin{bmatrix} I & 0 \\ 0 & 0 \end{bmatrix} \tilde{N}, \quad A = \tilde{M} \begin{bmatrix} \tilde{A} & 0 \\ 0 & I \end{bmatrix} \tilde{N}. \tag{2.69}$$

In this case, S can be written as

$$S = \tilde{M}^{-T} \begin{bmatrix} 0 \\ I \end{bmatrix} \tilde{H}, \tag{2.70}$$

where $\tilde{H}$ is any nonsingular matrix. Write

$$\tilde{M}^T P \tilde{M} = \begin{bmatrix} \tilde{P}_1 & \tilde{P}_2 \\ \tilde{P}_2^T & \tilde{P}_3 \end{bmatrix}, \quad \tilde{N}^{-T} Q \tilde{H}^T = \begin{bmatrix} \tilde{Q}_1 \\ \tilde{Q}_2 \end{bmatrix}, \tag{2.71}$$

where the partition is compatible with that of A in (2.69). Now, substituting (2.69)–(2.71) into (2.59) yields

$$\tilde{N}^T \begin{bmatrix} \tilde{A}^T \tilde{P}_1 \tilde{A} - \tilde{P}_1 & \tilde{A}^T \tilde{P}_2 + \tilde{Q}_1 \\ \tilde{P}_2^T \tilde{A} + \tilde{Q}_1^T & \tilde{P}_3 + \tilde{Q}_2 + \tilde{Q}_2^T \end{bmatrix} \tilde{N} < 0$$

which implies

$$\tilde{A} \tilde{P}_1 \tilde{A}^T - \tilde{P}_1 < 0.$$

Noting this and $\tilde{P}_1 > 0$, we have that $\rho(\tilde{A}) < 1$. Therefore, the pair (E, A) is stable. This together with the regularity and causality of the pair (E, A) gives that the pair (E, A) is admissible. This completes the proof.   □

Remark 2.5. In the case when $E = I$, that is, the unforced discrete singular system in (2.36) reduces to a state-space system, it is easy to see that $S = 0$ and Theorem 2.4 coincides with the Lyapunov stability theory in [4]. Therefore, Theorem 2.2 can be regarded as an extension of the Lyapunov stability theory for discrete state-space systems to discrete singular systems.   ◁

To illustrate the result in Theorem 2.4, we present the following numerical example.

Example 2.2. Consider an unforced discrete singular system in (2.36) with

$$E = \begin{bmatrix} 3 & 0 & 2 & -5 \\ 0 & 3 & -2 & 2 \\ 2 & 2 & 0 & -2 \\ 2 & -4 & 4 & -6 \end{bmatrix}, \quad A = \begin{bmatrix} 0.7 & -3.25 & -0.7 & 0 \\ 1.8 & 0.4 & -6.4 & 2.6 \\ 1 & -1.9 & -5.4 & 2.4 \\ -0.6 & -2.7 & 5.4 & -2.8 \end{bmatrix}.$$

It is easy to verify that rank$(E) = 2$. In order to use Theorem 2.4 to check the admissibility of this singular system, we choose

$$S = \begin{bmatrix} 0 & 1 \\ 1 & 1 \\ -0.5 & -1.5 \\ 0.5 & 0 \end{bmatrix},$$

which is with full column rank and satisfies $E^T S = 0$. Then, we obtain a set of solutions to the LMI in (2.59) as follows:

$$P = \begin{bmatrix} 0.6630 & -0.0013 & -0.2976 & -0.2949 \\ -0.0013 & 0.8292 & -0.1868 & 0.3709 \\ -0.2976 & -0.1868 & 0.7852 & 0.0507 \\ -0.2949 & 0.3709 & 0.0507 & 0.4169 \end{bmatrix},$$

$$Q = \begin{bmatrix} 4.8145 & -4.4723 \\ 6.0212 & -1.7657 \\ 15.3479 & 1.8695 \\ -11.2932 & -0.9923 \end{bmatrix}.$$

Therefore, by Theorem 2.4, it is easy to see that the singular system in this example is admissible. This can also be verified by Lemmas 2.9 and 2.10. To show this, we choose two nonsingular matrices

$$M = \begin{bmatrix} 3 & 2 & 2 & 0 \\ 0 & -2 & 4 & -2 \\ 2 & 0 & 4 & -2 \\ 2 & 4 & -2 & 2 \end{bmatrix}, \quad N = \begin{bmatrix} 1 & 1 & 0 & -1 \\ 0 & -1.5 & 1 & -1 \\ 1 & 0 & -1 & 0 \\ 1 & 0 & 1 & -1 \end{bmatrix}.$$

Then, it can be seen that

$$E = M \begin{bmatrix} I_2 & 0 \\ 0 & 0 \end{bmatrix} N, \quad A = M \begin{bmatrix} \mathcal{A} & 0 \\ 0 & I_2 \end{bmatrix} N,$$

where

$$\mathcal{A} = \begin{bmatrix} -0.5 & 0.3 \\ 0.1 & 0.2 \end{bmatrix}.$$

It can be verified that the two eigenvalues of the matrix $\mathcal{A}$ are -0.5405 and 0.2405; that is, $\rho(\mathcal{A}) < 1$. Hence, by Lemmas 2.9 and 2.10, we have that the singular system is admissible. ◇

2.4 Conclusion

This chapter has addressed the stability of singular systems in both the continuous and discrete cases. Without decomposing the system matrices, we have obtained necessary and sufficient conditions which guarantee a given singular system not only asymptotically stable but also regular and impulse-free (in the continuous case) or causal (in the discrete case). The conditions are expressed in terms of LMIs, which can be efficiently handled by using standard numerical algorithms. It is worth mentioning that the stability results developed in this chapter will play important roles in dealing with the stabilization problem as can be seen in the next chapter. Part of the results presented in this chapter have also appeared in [179, 189].

3

Stabilization

3.1 Introduction

In this chapter, we shall deal with the stabilization problem for singular systems. The purpose is the design of controllers such that the closed-loop system is regular, stable and impulse-free (in the continuous case) or causal (in the discrete case). For continuous singular systems, both state feedback and dynamic output feedback controllers are considered. Based on the stability conditions presented in Chapter 2, the stabilizing controller design can be formulated as a convex optimization problem characterized by linear matrix inequalities (LMI). In the context of discrete singular systems, a necessary and sufficient condition for the existence of stabilizing state feedback controllers is obtained.

3.2 Continuous Systems

In this section, we first consider the design of a state feedback controller for continuous singular systems such that the closed-loop system is admissible. Then the design of dynamic output feedback controllers is investigated. Necessary and sufficient conditions are given and an LMI approach is developed to obtain desired feedback controllers.

3.2.1 State Feedback Control

Here, we assume that all the state variables are available for state feedback. The purpose is the design of state feedback controllers for continuous singular systems such that the closed-loop system is admissible.

Consider the following linear continuous singular system:

$$E\dot{x}(t) = Ax(t) + Bu(t), \tag{3.1}$$

where $x(t) \in \mathbb{R}^n$ is the state; $u(t) \in \mathbb{R}^m$ is the control input. The matrix $E \in \mathbb{R}^{n \times n}$ may be singular; we shall assume that $\text{rank}(E) = r \leq n$. A and B are known real constant matrices with appropriate dimensions. For the continuous singular system in (3.1), we consider the following state feedback controller:

$$u(t) = Kx(t), \quad K \in \mathbb{R}^{m \times n}. \tag{3.2}$$

Applying this controller to system in (3.1), we obtain the closed-loop system as follows:

$$E\dot{x}(t) = (A + BK)x(t). \tag{3.3}$$

Then, we have the following stabilization result.

Theorem 3.1. *Consider the continuous singular system (3.1). There exists a state feedback controller (3.2) such that the closed-loop system (3.3) is admissible if and only if there exist matrices $P > 0$, Q and Y such that*

$$\Omega(P,Q)^T A^T + A\Omega(P,Q) + BY + Y^T B^T < 0 \tag{3.4}$$

where

$$\Omega(P,Q) = PE^T + SQ, \tag{3.5}$$

and $S \in \mathbb{R}^{n \times (n-r)}$ is any matrix with full column rank and satisfies $ES = 0$. In this case, we can assume that the matrix $\Omega(P,Q)$ is nonsingular (if this is not the case, then we can choose some $\theta \in (0,1)$ such that $\hat{\Omega}(P,Q) = \Omega(P,Q) + \theta\tilde{P}$ is nonsingular and satisfies (3.4), in which $\tilde{P}$ is any nonsingular matrix satisfying $E\tilde{P} = \tilde{P}^T E^T \geq 0$), then a stabilizing state feedback controller can be chosen as

$$u(t) = Y\Omega(P,Q)^{-1}x(t). \tag{3.6}$$

Proof. (*Sufficiency*) Applying the state feedback controller in (3.6) to (3.1) results in the closed-loop system as

$$E\dot{x}(t) = \left(A + BY\Omega(P,Q)^{-1}\right)x(t). \tag{3.7}$$

It is easy to see that (3.4) can be rewritten as

$$\Omega(P,Q)^T \left(A + BY\Omega(P,Q)^{-1}\right)^T + \left(A + BY\Omega(P,Q)^{-1}\right)\Omega(P,Q) < 0.$$

Noting the definition of the matrix S and using Corollary 2.2, we have that the closed-loop system (3.7) is admissible.

(*Necessity*) Assume that there exists a state feedback controller (3.2) such that the closed-loop system (3.3) is admissible. Then, it follows from Corollary 2.2 that there exist matrices $P > 0$ and Q such that

$$\left(PE^T + SQ\right)^T (A + BK)^T + (A + BK)\left(PE^T + SQ\right) < 0.$$

Set

$$Y = K\left(PE^T + SQ\right).$$

Then, the desired result follows immediately.   □

To show the effectiveness of the result in Theorem 3.1, we provide the following numerical example.

Example 3.1. Consider a continuous singular system in (3.1) with parameters as follows:

$$E = \begin{bmatrix} 1 & 1 & -1 \\ -0.5 & -0.5 & -0.5 \\ 1 & 1 & -1 \end{bmatrix}, \quad A = \begin{bmatrix} 3 & 4 & -2.8 \\ -2 & -1 & 0.1 \\ 5 & 4 & -4.8 \end{bmatrix},$$

$$B = \begin{bmatrix} 1 & -1 & 7 \\ -4.5 & -0.5 & 3 \\ 1 & 3 & 9 \end{bmatrix}.$$

It can be verified that $\text{rank}(E) = 2$, and there are two finite eigenvalues of the pair (E, A) given by 3.9050 and 1.8950. Thus, the open-loop system is not stable, and hence not admissible. To construct a stabilizing state feedback controller for this system, we first choose

$$S = \begin{bmatrix} 1 \\ -1 \\ 0 \end{bmatrix},$$

which is with full column rank and satisfies $ES = 0$. Then, it can be found that the LMI in (3.4) is feasible. Therefore, by Theorem 3.1, we can choose a state feedback controller

$$u(t) = \begin{bmatrix} -4.7518 & 2.7703 & -0.9636 \\ 1.8541 & -1.6544 & 0.1983 \\ -0.1417 & -1.1694 & 0.0368 \end{bmatrix} x(t), \tag{3.8}$$

such that the closed-loop system is admissible.

To verify the admissibility of the closed-loop system, we apply (3.8) to the above system and obtain the closed-loop system as

$$E\dot{x}(t) = A_c x(t), \tag{3.9}$$

where

$$A_c = \begin{bmatrix} -4.5976 & 0.2389 & -3.7044 \\ 18.0312 & -16.1473 & 4.4477 \\ 4.5355 & -8.7177 & -4.8378 \end{bmatrix}.$$

Now, choose two nonsingular matrices

$$M = \begin{bmatrix} 0.7326 & 0 & 0.2674 \\ -1.5231 & -1 & 2.0231 \\ -0.5000 & 0 & 0.5000 \end{bmatrix},$$

$$N = \begin{bmatrix} -0.0627 & 0.5578 & 0.1106 \\ 0.0627 & 0.4422 & -0.1106 \\ -1 & 1 & 0 \end{bmatrix}.$$

Then, it is easy to verify that

$$MEN = \begin{bmatrix} I_2 & 0 \\ 0 & 0 \end{bmatrix}, \quad MAN = \begin{bmatrix} \mathcal{A}_c & 0 \\ 0 & 1 \end{bmatrix},$$

where

$$\mathcal{A}_c = \begin{bmatrix} 4.0074 & -6.1632 \\ 8.5928 & -10.4460 \end{bmatrix}.$$

It can be calculated that the two eigenvalues of the matrix $\mathcal{A}_c$ are $-3.2193 + 0.8566i$ and $-3.2193 - 0.8566i$; that is, $\alpha(\mathcal{A}_c) < 0$. Therefore, by Lemma 2.2, the closed-loop system (3.9) is admissible.   ◇

3.2.2 Output Feedback Control

In practical applications, usually not all the state variables are available for feedback. Here we will design dynamic output feedback controllers for continuous singular systems such that the closed-loop system is admissible.

Consider a linear continuous singular system described by

$$E\dot{x}(t) = Ax(t) + Bu(t), \tag{3.10}$$
$$y(t) = Cx(t) + Du(t), \tag{3.11}$$

where $x(t) \in \mathbb{R}^n$ is the state; $u(t) \in \mathbb{R}^m$ is the control input; $y(t) \in \mathbb{R}^q$ is the measurement. The matrix $E \in \mathbb{R}^{n \times n}$ may be singular; we shall assume that $\text{rank}(E) = r \leq n$. A, B, C and D are known real constant matrices with appropriate dimensions.

Now, we consider the following dynamic output feedback controller:

$$E_K \dot{\xi}(t) = A_K \xi(t) + B_K y(t), \tag{3.12}$$

$$u(t) = C_K \xi(t), \tag{3.13}$$

where $\xi(t) \in \mathbb{R}^n$ is the controller state, E_K, A_K, B_K and C_K are constant matrices to be determined. Applying this controller to the singular system in (3.10) and (3.11) results in the following closed-loop system:

$$E_c \eta(t) = A_c \eta(t), \tag{3.14}$$

where

$$\eta(t) = \begin{bmatrix} x(t) \\ \xi(t) \end{bmatrix}, \tag{3.15}$$

and

$$E_c = \begin{bmatrix} E & 0 \\ 0 & E_K \end{bmatrix}, \quad A_c = \begin{bmatrix} A & BC_K \\ B_K C & A_K + B_K DC_K \end{bmatrix}. \tag{3.16}$$

Before proceeding further, we introduce the following lemma, which will be used in the sequel.

Lemma 3.1. [15] *The matrix inequality*

$$\begin{bmatrix} Z_1 & Z_2 \\ Z_2^T & Z_3 \end{bmatrix} \geq 0,$$

holds if and only if

$$Z_3 \geq 0,$$
$$Z_1 - Z_2 Z_3^+ Z_2^T \geq 0,$$
$$Z_2 \left(I - Z_3 Z_3^+ \right) = 0.$$

Now, we are in a position to present the output feedback stabilization result in the following theorem.

Theorem 3.2. *Consider the continuous singular system in (3.10) and (3.11). There exists a dynamic output state feedback controller in the form of (3.12) and (3.13) such that the closed-loop system (3.14) is admissible if and only if there exist matrices X, Y, Φ, and Ψ such that*

$$\begin{bmatrix} E^T & 0 \\ 0 & E \end{bmatrix} \begin{bmatrix} X & I \\ I & Y \end{bmatrix} = \begin{bmatrix} X^T & I \\ I & Y^T \end{bmatrix} \begin{bmatrix} E & 0 \\ 0 & E^T \end{bmatrix} \geq 0, \tag{3.17}$$

$$A^T X + X^T A + \Phi C + C^T \Phi^T < 0, \tag{3.18}$$

$$AY + Y^T A^T - B\Psi - \Psi^T B^T < 0. \tag{3.19}$$

If (3.17)–(3.19) hold, then we can always find matrices X, Y, Φ and Ψ that satisfy (3.17)–(3.19) and both Y and $Y^{-1} - X$ are nonsingular, and a desired stabilizing dynamic output feedback controller in (3.12) and (3.13) can be chosen with the following parameters:

$$A_K = \left(X - Y^{-1}\right)^{-T}\left(A^T Y^{-1} + X^T A - X^T B\Psi Y^{-1} + \Phi C\right)$$
$$+ \left(Y^{-1} - X\right)^{-T}\Phi D\Psi Y^{-1}, \tag{3.20}$$
$$E_K = E, \tag{3.21}$$
$$B_K = \left(Y^{-1} - X\right)^{-T}\Phi, \tag{3.22}$$
$$C_K = -\Psi Y^{-1}. \tag{3.23}$$

Proof. (*Sufficiency*) When (3.17)–(3.19) are satisfied, we first show that there always exist matrices X, Y, Φ and Ψ that satisfy (3.17)–(3.19) and both Y and $Y^{-1} - X$ are nonsingular. To this end, we note that (3.17) implies

$$E^T X = X^T E \geq 0, \tag{3.24}$$
$$EY = Y^T E^T \geq 0. \tag{3.25}$$

If Y is singular, then we choose any nonsingular matrix $\tilde{Y}$ satisfying

$$E\tilde{Y} = \tilde{Y}^T E^T \geq 0.$$

Let $\theta_1 \in (0, 1)$ satisfy that θ_1 is not an eigenvalue of $-Y\tilde{Y}^{-1}$ and is small enough such that when Y is replaced by $Y + \theta_1\tilde{Y}$, the inequalities in (3.17)–(3.19) still hold. Then, it is easy to see that $Y + \theta_1\tilde{Y}$ is nonsingular. Now, noting (3.25) and applying Lemma 3.1 to (3.17), we have

$$E^T X - E^T (EY)^+ E$$
$$= E^T X - Y^{-T}\left(Y^T E^T\right)(EY)^+ E$$
$$= E^T X - Y^{-T}(EY)(EY)^+(EY)Y^{-1}$$
$$= E^T X - Y^{-T}(EY)Y^{-1}$$
$$= E^T X - Y^{-T}E$$
$$= E^T\left(X - Y^{-1}\right)$$
$$= \left(X - Y^{-1}\right)^T E$$
$$\geq 0. \tag{3.26}$$

If $Y^{-1} - X$ is singular, then we choose any nonsingular matrix $\tilde{X}$ satisfying

$$E^T\tilde{X} = \tilde{X}^T E \geq 0.$$

Let $\theta_2 \in (0, 1)$ satisfy that θ_2 is not an eigenvalue of $\left(Y^{-1} - X\right)\tilde{X}^{-1}$ and is small enough such that when X is replaced by $X + \theta_2\tilde{X}$, the inequalities in (3.17)–(3.19) still hold. Then, it is easy to see that $\left(X + \theta_2 \tilde{X}\right) - Y^{-1}$ is nonsingular and satisfies (3.26). Now, by the output feedback controller with the parameters given in (3.20)–(3.23), we have the closed-loop system as

$$\tilde{E}_c \eta(t) = \tilde{A}_c \eta(t), \tag{3.27}$$

where $\eta(t)$ is given in (3.15), and

$$\tilde{E}_c = \begin{bmatrix} E & 0 \\ 0 & E \end{bmatrix}, \tag{3.28}$$

$$\tilde{A}_c = \begin{bmatrix} A & -B\Psi Y^{-1} \\ \left(Y^{-1} - X\right)^{-T} \Phi C & \tilde{A} \end{bmatrix}, \tag{3.29}$$

where

$$\tilde{A} = \left(X - Y^{-1}\right)^{-T} \left(A^T Y^{-1} + X^T A - X^T B\Psi Y^{-1} + \Phi C\right).$$

Set

$$\tilde{P}_c = \begin{bmatrix} X & Y^{-1} - X \\ Y^{-1} - X & X - Y^{-1} \end{bmatrix}.$$

Then, considering (3.25) and (3.26), it can be deduced that

$$\begin{aligned}
& E^T X - E^T \left(Y^{-1} - X\right) \left[E^T \left(X - Y^{-1}\right)\right]^+ \left[E^T \left(Y^{-1} - X\right)\right] \\
&= E^T X - E^T \left(X - Y^{-1}\right) \\
&= E^T Y^{-1} \\
&\geq 0, \tag{3.30}
\end{aligned}$$

and

$$\begin{aligned}
& E^T \left(Y^{-1} - X\right) \left[I - E^T \left(X - Y^{-1}\right) \left[E^T \left(X - Y^{-1}\right)\right]^+\right] \\
&= E^T \left(Y^{-1} - X\right) + E^T \left(X - Y^{-1}\right) \left[E^T \left(X - Y^{-1}\right)\right]^T \\
&\quad \times \left[\left(E^T \left(X - Y^{-1}\right)\right)^T\right]^+ \\
&= E^T \left(Y^{-1} - X\right) + E^T \left(X - Y^{-1}\right) \\
&\quad \times \left[\left(E^T \left(X - Y^{-1}\right)\right)^+ \left(E^T \left(X - Y^{-1}\right)\right)\right]^T \\
&= E^T \left(Y^{-1} - X\right) + E^T \left(X - Y^{-1}\right) \\
&\quad \times \left[E^T \left(X - Y^{-1}\right)\right]^+ \left(E^T \left(X - Y^{-1}\right)\right) \\
&= E^T \left(Y^{-1} - X\right) + E^T \left(X - Y^{-1}\right) \\
&= 0. \tag{3.31}
\end{aligned}$$

Then, noting (3.24), (3.30) and (3.31), and using Lemma 3.1, we have

$$\tilde{E}_c^T \tilde{P}_c = \tilde{P}_c^T \tilde{E}_c \geq 0. \tag{3.32}$$

Now, it can be verified via algebraic manipulations that

$$\tilde{A}_c^T \tilde{P}_c + \tilde{P}_c^T \tilde{A}_c = \begin{bmatrix} \Gamma_{11} & -\Gamma_{11} \\ \Gamma_{11} & \Gamma_{22} \end{bmatrix},$$

where

$$\Gamma_{11} = A^T X + X^T A + \Phi C + C^T \Phi^T,$$
$$\Gamma_{22} = A^T X + X^T A + \Phi C + C^T \Phi^T$$
$$+ Y^{-T} A + A^T Y^{-1} - Y^{-T} \left(B\Psi - \Psi^T B^T \right) Y^{-1}.$$

On the other hand, pre- and post-multiplying (3.19) by Y^{-T} and Y^{-1}, respectively, give

$$Y^{-T} A + A^T Y^{-1} - Y^{-T} \left(B\Psi - \Psi^T B^T \right) Y^{-1} < 0.$$

This together with (3.18) implies

$$\Gamma_{22} < 0.$$

Note $\Gamma_{11} < 0$ and

$$\Gamma_{22} - \Gamma_{11} < 0.$$

Then, by Schur complement, we have

$$\begin{bmatrix} \Gamma_{11} & \Gamma_{11} \\ \Gamma_{11} & \Gamma_{22} \end{bmatrix} < 0.$$

That is,

$$\tilde{A}_c^T \tilde{P}_c + \tilde{P}_c^T \tilde{A}_c < 0. \tag{3.33}$$

Therefore, by (3.32) and (3.33) and Theorem 2.1, we have that the closed-loop system (3.27) is admissible.

(*Necessity*) Assume that there exists a dynamic output feedback controller in (3.12) and (3.13) such that the closed-loop system (3.14) is admissible. Then, it follows from Theorem 2.1 that there exists a matrix P_c such that

$$E_c^T P_c = P_c^T E_c \geq 0, \tag{3.34}$$
$$P_c^T A_c + A_c^T P_c < 0. \tag{3.35}$$

Write

$$P_c = \begin{bmatrix} P_{c1} & P_{c2} \\ P_{c3} & P_{c4} \end{bmatrix},$$

where the partition is compatible with that for A_c and E_c in (3.16). Then, it follows from (3.34) that

$$E^T P_{c1} = P_{c1}^T E \geq 0, \tag{3.36}$$

$$E_K^T P_{c4} = P_{c4}^T E_K \geq 0, \tag{3.37}$$

$$E^T P_{c2} = P_{c3}^T E_K. \tag{3.38}$$

Without loss of generality, we can assume P_{c4} is nonsingular. If not, from (3.37), we choose any nonsingular matrix $\tilde{P}_{c4}$ satisfying

$$E_K^T \tilde{P}_{c4} = \tilde{P}_{c4}^T E_K \geq 0.$$

Let $\alpha_1 \in (0,1)$ satisfy that α_1 is not an eigenvalue of $-P_{c4}\tilde{P}_{c4}^{-1}$ and is small enough such that when P_c is replaced by

$$P_c + \alpha_1 \begin{bmatrix} 0 & 0 \\ 0 & \tilde{P}_{c4} \end{bmatrix}$$

the inequality in (3.35) still holds. Then, it is easy to see that $P_{c4} + \alpha_1 \tilde{P}_{c4}$ is nonsingular. Similarly, without loss of generality, we can assume $P_{1c} - P_{2c}P_{4c}^{-1}P_{3c}$ is nonsingular. Now, by the 1-1 block of (3.35), we have

$$A^T P_{1c} + P_{1c}^T A + P_{3c}^T B_K C + C^T B_K^T P_{3c} < 0. \tag{3.39}$$

Set

$$X = P_{c1}, \quad \Phi = P_{3c}^T B_K. \tag{3.40}$$

Then, it is easy to see that (3.39) provides (3.18), and (3.36) provides

$$E^T X = X^T E \geq 0. \tag{3.41}$$

Now, let

$$U = \begin{bmatrix} I & 0 \\ -P_{4c}^{-1} P_{3c} & I \end{bmatrix}.$$

Pre- and post-multiplying (3.35) by U^T and U, respectively, and then noting the 1-1 block, we obtain

$$\left(P_{1c} - P_{2c}P_{4c}^{-1}P_{3c}\right)^T A + A^T \left(P_{1c} - P_{2c}P_{4c}^{-1}P_{3c}\right)$$
$$- \left(P_{1c} - P_{2c}P_{4c}^{-1}P_{3c}\right)^T BC_K P_{4c}^{-1} P_{3c}$$
$$- P_{3c}^T P_{4c}^{-T} C_K^T B^T \left(P_{1c} - P_{2c}P_{4c}^{-1}P_{3c}\right) < 0. \tag{3.42}$$

Furthermore, the 1-1 block of

$$U^T E_c^T P_c U = U^T P_c^T E_c U \geq 0,$$

gives

$$E^T \left(P_{1c} - P_{2c}P_{4c}^{-1}P_{3c}\right) = \left(P_{1c} - P_{2c}P_{4c}^{-1}P_{3c}\right)^T E \geq 0. \tag{3.43}$$

Let

$$Y = \left(P_{1c} - P_{2c}P_{4c}^{-1}P_{3c}\right)^{-1}, \tag{3.44}$$

$$\Psi = C_K P_{4c}^{-1} P_{3c} Y. \tag{3.45}$$

Then, pre- and post-multiplying (3.42) by Y^T and Y, respectively, we have that (3.19) holds. From (3.37) and (3.43), it can be shown that

$$P_{c4}^{-T} E_K^T = E_K P_{c4}^{-1} \geq 0,$$

and

$$Y^T E^T = EY \geq 0. \tag{3.46}$$

Therefore

$$
\begin{aligned}
&E^T \left(X - Y^{-1}\right)\\
&= E^T \left[P_{c1} - \left(P_{1c} - P_{2c}P_{4c}^{-1}P_{3c}\right)\right]\\
&= E^T P_{2c} P_{4c}^{-1} P_{3c}\\
&= P_{c3}^T E_K P_{4c}^{-1} P_{3c}\\
&= P_{c3}^T P_{c4}^{-T} E_K^T P_{3c}\\
&= E^T \left(X - Y^{-1}\right)^T\\
&\geq 0. \tag{3.47}
\end{aligned}
$$

Observe that

$$
\begin{aligned}
&E^T X - E^T \left(EY\right)^+ E\\
&= E^T X - Y^{-T} \left(EY\right)\left(EY\right)^+ \left(EY\right) Y^{-1}\\
&= E^T X - Y^{-T} \left(EY\right) Y^{-1}\\
&= E^T X - Y^{-T} E\\
&= E^T \left(X - Y^{-1}\right)\\
&\geq 0. \tag{3.48}
\end{aligned}
$$

and

$$
\begin{aligned}
&E^T \left[I - EY\left(EY\right)^+\right]\\
&= E^T - E^T \left[\left(EY\right)\left(EY\right)^+\right]^T\\
&= E^T - Y^{-T}\left(Y^T E^T\right)\left(Y^T E^T\right)^+ \left(Y^T E^T\right)\\
&= E^T - Y^{-T}\left(Y^T E^T\right)\\
&= 0. \tag{3.49}
\end{aligned}
$$

Noting (3.41), (3.47), (3.48) and (3.49) and using Lemma 3.1, we have that (3.17) is satisfied. This completes the proof. $\qquad\square$

3.3 Discrete Systems

In this section, we present results on the stabilization for discrete singular systems. The class of linear discrete singular systems to be considered is described by

$$Ex(k+1) = Ax(k) + Bu(k), \tag{3.50}$$

where $x(k) \in \mathbb{R}^n$ is the state; $u(k) \in \mathbb{R}^m$ is the control input. The matrix $E \in \mathbb{R}^{n \times n}$ may be singular; we shall assume that $\text{rank}(E) = r \leq n$. A and B are known real constant matrices with appropriate dimensions. For the discrete-time singular system in (3.50), we assume that all the state variables are available for state feedback, and consider the following state feedback controller:

$$u(k) = Kx(k), \quad K \in \mathbb{R}^{m \times n}. \tag{3.51}$$

Applying this controller to system in (3.50) results in the following closed-loop system:

$$Ex(k+1) = (A + BK)x(k). \tag{3.52}$$

The purpose of the stabilization problem is the design of a state feedback controller in (3.51) such that the closed-loop system (3.52) is admissible.

Theorem 3.3. *Consider the discrete-time singular system in (3.50). There exists a state feedback controller (3.51) such that the closed-loop system (3.52) is admissible if and only if there exist a scalar $\delta > 0$, matrices $P > 0$ and Q such that*

$$QS^T A + A^T SQ^T - E^T PE + A^T PA$$
$$- \left(QS^T + A^T P\right) B \left(B^T PB + \delta I\right)^{-1} B^T \left(QS^T + A^T P\right)^T < 0, \tag{3.53}$$

where $S \in \mathbb{R}^{n \times (n-r)}$ is any matrix with full column rank and satisfies $E^T S = 0$. In this case, a desired stabilizing state feedback control law is given by

$$u(k) = - \left(B^T PB + \delta I\right)^{-1} B^T \left(QS^T + A^T P\right)^T x(k). \tag{3.54}$$

Proof. (*Sufficiency*) Applying the state feedback controller in (3.54) to (3.50) results in the closed-loop system as

$$Ex(k+1) = (A + BK_c)x(k), \tag{3.55}$$

where

$$K_c = - \left(B^T PB + \delta I\right)^{-1} B^T \left(QS^T + A^T P\right)^T.$$

By some algebraic manipulations, it can be verified that

$$\left(QS^T + A^T P\right) BK_c + K_c^T B^T \left(QS^T + A^T P\right)^T + K^T \left(B^T PB + \delta I\right) K$$
$$= -\left(QS^T + A^T P\right) B \left(B^T PB + \delta I\right)^{-1} B^T \left(QS^T + A^T P\right)^T.$$

Then

$$(A + BK_c)^T P (A + BK_c) - E^T PE + QS^T (A + BK_c)$$
$$+ (A + BK_c)^T SQ^T$$
$$\leq A^T PA - E^T PE + QS^T A + A^T SQ^T + \left(QS^T + A^T P\right) BK_c$$
$$+ K_c^T B^T \left(QS^T + A^T P\right)^T + K_c^T \left(B^T PB + \delta I\right) K_c$$
$$= QS^T A + A^T SQ^T - E^T PE + A^T PA$$
$$- \left(QS^T + A^T P\right) B \left(B^T PB + \delta I\right)^{-1} B^T \left(QS^T + A^T P\right)^T$$
$$< 0,$$

which, by Theorem 2.4, implies that the closed-loop system (3.55) is admissible.

(*Necessity*) Assume that there exists a state feedback controller (3.51) such that the closed-loop system (3.52) is admissible. Then, it follows from Theorem 2.4 that there exist matrices $P > 0$ and Q such that

$$(A + BK)^T P (A + BK) - E^T PE$$
$$+ QS^T (A + BK) + (A + BK)^T SQ^T < 0. \tag{3.56}$$

That is,

$$A^T PA - E^T PE + QS^T A + A^T SQ^T + \left(QS^T + A^T P\right) BK$$
$$+ K^T B^T \left(QS^T + A^T P\right)^T + K^T B^T PBK < 0,$$

which implies that there exists a scalar $\delta > 0$ such that

$$A^T PA - E^T PE + QS^T A + A^T SQ^T + \left(QS^T + A^T P\right) BK$$
$$+ K^T B^T \left(QS^T + A^T P\right)^T + K^T \left(B^T PB + \delta I\right) K < 0.$$

This can be rewritten as

$$A^T PA - E^T PE + QS^T A + A^T SQ^T$$
$$- \left(QS^T + A^T P\right) B \left(B^T PB + \delta I\right)^{-1} B^T \left(QS^T + A^T P\right)^T$$
$$+ \left[K^T + \left(QS^T + A^T P\right) B \left(B^T PB + \delta I\right)^{-1}\right] \left(B^T PB + \delta I\right)$$
$$\times \left[K^T + \left(QS^T + A^T P\right) B \left(B^T PB + \delta I\right)^{-1}\right]^T < 0,$$

which implies that (3.53) is satisfied. This completes the proof. $\square$

Remark 3.1. Theorem 3.3 provides a necessary and sufficient condition for the existence of stabilizing state feedback controllers for discrete singular systems. This result is theoretically important and mathematically elegant considering that many existing necessary and sufficient conditions for state feedback stabilization results are obtained under some assumptions on the matrix B [189, 192]. ◁

Remark 3.2. It is noted that solving the matrix inequality in (3.53) is not an easy task. However, from the proof of Theorem 3.3, it can be found that (3.56) is equivalent to

$$\left[\begin{array}{cc} QS^T (A + BK) + (A + BK)^T SQ^T - E^T PE & (A + BK)^T P \\ P (A + BK) & -P \end{array} \right] < 0. \quad (3.57)$$

This is a bilinear matrix inequality (BMI). Therefore, to design desired state feedback controllers, we can resort to solving (3.57) rather than solving (3.53) directly. For this purpose, many existing effective algorithms, such as the branch-and-bound algorithm, the Lagrangian dual global optimization algorithm and the branch-and-cut algorithm proposed in [160], [161] and [54], respectively, can be used. ◁

Now, we provide the following illustrative example.

Example 3.2. Consider a discrete singular system in (3.50) with

$$E = \left[\begin{array}{ccc} 1 & 3 & 0 \\ 2 & -1 & 0 \\ 1 & 5 & 0 \end{array} \right], \quad A = \left[\begin{array}{ccc} 4.6 & -2.8 & 0 \\ 2.2 & 2.8 & 0 \\ 6.6 & -5.2 & 0 \end{array} \right],$$

$$B = \left[\begin{array}{ccc} 1.9 & -4 & -1.2 \\ 1.7 & 1 & -1.4 \\ 2.5 & -2 & 0.8 \end{array} \right].$$

It can be verified that this system is neither regular nor stable. To construct a stabilizing state feedback controller, we choose

$$S = \left[\begin{array}{c} -0.6875 \\ 0.1250 \\ 0.4375 \end{array} \right],$$

which is with full column rank and satisfies $E^T S = 0$. Then, it can be found that a set of solutions to (3.57) is as follows:

$$P = \left[\begin{array}{ccc} 0.5492 & 0.3193 & -0.2747 \\ 0.3193 & 0.3208 & -0.2368 \\ -0.2747 & -0.2368 & 0.1946 \end{array} \right], \quad (3.58)$$

$$Q = \begin{bmatrix} -0.5883 \\ 2.1832 \\ -0.1364 \end{bmatrix}, \quad K = \begin{bmatrix} -1.6476 & -1.9411 & 0.2247 \\ 0.5181 & -1.8505 & 0.0734 \\ -0.5582 & 1.0690 & 0.0841 \end{bmatrix}. \quad (3.59)$$

Then, with the state feedback gain K given in (3.59), it is easy to verify that the closed-loop system is admissible. ◇

3.4 Conclusion

This chapter has studied the stabilization problem for both continuous and discrete singular systems. An LMI approach has been developed to design both stabilizing state feedback and dynamic output feedback controllers for continuous singular systems, while in the discrete case, a BMI approach has been presented to design stabilizing state feedback controllers. It is worth noting that in both the continuous and discrete cases, the solvability conditions of the stabilization problem have been given without decomposing the system matrices.

4

Robust Control

4.1 Introduction

The problems of robust stability and robust stabilization of linear state-space systems with parameter uncertainties have been extensively studied in the past decades [25, 209]. Many results on these topics have been proposed. Among the different approaches that deal with these problems, the methods based on the concepts of quadratic stability and quadratic stabilizability have become popular. An uncertain system is quadratically stable if there exists a fixed Lyapunov function to infer the stability of the uncertain system, while an uncertain system is quadratically stabilizable if there exists a feedback controller such that the closed-loop system is quadratically stable. Many results on quadratic stability and quadratic stabilizability have been reported in both the continuous and discrete contexts; see, e.g., [8, 57, 182, 210], and the references therein. However, in the context of uncertain singular systems, it has been shown that the problems of robust stability and robust stabilization are more complicated than those for state-space systems. Specifically, the robust stability problem for singular systems requires considering not only stability robustness, but also regularity and non-impulsiveness (for continuous singular systems) or causality (for discrete singular systems) [48, 49, 112, 177, 183], while the latter two are intrinsic properties of state-space systems. Similarly, the robust stabilization problem for uncertain singular systems must consider not only stabilization but also regularization and impulse elimination, while the latter two issues do not arise in the state-space case.

In this chapter, we consider the problems of robust stability analysis and robust stabilization for uncertain singular systems. The concepts of generalized quadratic stability and generalized quadratic stabilizability for both continuous- and discrete-time singular systems are proposed. It is shown that generalized quadratic stability and generalized quadratic stabilizability imply robust stability and robust stabilizability, respectively. In the case when the parameter uncertainties are time-invariant and norm-bounded, necessary and sufficient conditions for both generalized quadratic stability and generalized quadratic stabilizability are obtained. When the parameter uncertainties are time-invariant and belong to a convex bounded domain (polytopic type uncertainty), robust stability conditions are proposed via parameter-independent and parameter-dependent Lyapunov matrices, respectively. Necessary and sufficient conditions for generalized quadratic stabilizability are derived for both continuous and discrete singular systems with time-invariant norm-bounded parameter uncertainties. When these conditions are feasible, desired robustly stabilizing state feedback controllers are also constructed.

4.2 Robust Stability

In this section, we consider the robust stability problem for uncertain singular systems. The concept of generalized quadratic stability for both continuous and discrete singular systems will be proposed. Both time-invariant norm-bounded and polytopic type parameter uncertainties will be considered.

4.2.1 Continuous Systems

Here, we first consider an uncertain linear continuous singular system with time-invariant norm-bounded parameter uncertainty described by

$$E\dot{x}(t) = (A + \Delta A)\, x(t), \tag{4.1}$$

where $x(t) \in \mathbb{R}^n$ is the state. The matrix $E \in \mathbb{R}^{n \times n}$ may be singular; we shall assume that $\text{rank}(E) = r \leq n$. A is a known real constant matrix with appropriate dimensions. ΔA is an unknown time-invariant matrix representing norm-bounded parameter uncertainties, which is assumed to be of the form

$$\Delta A = MF(\sigma)N_1, \tag{4.2}$$

where M and N_1 are known real constant matrices with appropriate dimensions. The uncertain matrix $F(\sigma)$ satisfies

$$F(\sigma)^T F(\sigma) \leq I, \tag{4.3}$$

where $\sigma \in \Theta$, and Θ is a compact set in $\mathbb{R}$. Furthermore, it is assumed that given any matrix $F : F^T F \leq I$, there exists a $\sigma \in \Theta$ such that $F = F(\sigma)$.

As has been pointed out, to investigate the robust stability problem for the uncertain singular system in (4.1), we have to consider not only stability robustness, but also regularity and non-impulsiveness of (4.1). In view of this, we first introduce the following definition.

Definition 4.1. *The uncertain singular system in (4.1) is said to be robustly stable if it is regular, impulse-free and stable for all allowable uncertainties.*

Here, the purpose is to develop conditions for robust stability of continuous singular systems. Considering that the concept of quadratic stability has played an important role in dealing with the robust stability problem for uncertain continuous state-space systems, we introduce the following definition for the uncertain continuous singular system in (4.1).

Definition 4.2. *The uncertain continuous singular system in (4.1) is said to be generalized quadratically stable if there exist matrices $P > 0$ and Q such that*

$$\left(PE^T + SQ\right)^T (A + \Delta A)^T + (A + \Delta A)\left(PE^T + SQ\right) < 0, \qquad (4.4)$$

for all allowable $F(\sigma)$, where $S \in \mathbb{R}^{n \times (n-r)}$ is any matrix with full column rank and satisfies $ES = 0$.

Remark 4.1. In the case when $E = I$, it is easy to see that generalized quadratic stability coincides with the notion of quadratic stability [88]. Therefore, generalized quadratic stability can be regarded as an extension of quadratic stability for state-space systems to singular systems.   ◁

The following lemma shows that generalized quadratic stability implies robust stability.

Lemma 4.1. *If the uncertain continuous singular system (4.1) is generalized quadratically stable, then it is robustly stable.*

Proof. Suppose that the uncertain continuous singular system (4.1) is generalized quadratically stable. Then, by Definition 4.2, we have that (4.4) holds, which, by Corollary 2.2, implies that (4.1) is regular, impulse-free and stable. Therefore, by Definition 4.1, it can be seen that the uncertain singular system (4.1) is robustly stable.   □

In view of Lemma 4.1, in the following, attention will be focused on the derivation of conditions for generalized quadratic stability. To this end, we introduce the following lemmas, which will be used in the sequel.

Lemma 4.2. [88] *Let $x \in \mathbb{R}^n$, Y and Z be given constant matrices with appropriate dimensions. Then*

$$\max\left\{ \left(x^T ZFYx\right)^2 : F^T F \leq I \right\} = \left(x^T ZZ^T x\right)\left(x^T Y^T Yx\right).$$

Lemma 4.3. [88] *Let $x \in \mathbb{R}^n$, $X \geq 0$, $Y \geq 0$ and $Z < 0$ be given constant matrices with appropriate dimensions. Furthermore, assume that*

$$\left(x^T Zx\right)^2 - 4\left(x^T Xx\right)\left(x^T Yx\right) > 0,$$

for all $x \neq 0$. Then, there exists a constant $\lambda > 0$ such that

$$\lambda^2 X + \lambda Z + Y < 0.$$

Now, we are in a position to present a necessary and sufficient condition for generalized quadratic stability.

Theorem 4.1. *The uncertain singular system in (4.1) is generalized quadratically stable if and only if there exist matrices $P > 0$, Q and a scalar $\epsilon > 0$ such that the following LMI holds:*

$$\begin{bmatrix} \left(PE^T + SQ\right)^T A^T + A\left(PE^T + SQ\right) + \epsilon MM^T & \left(PE^T + SQ\right)^T N_1^T \\ N_1\left(PE^T + SQ\right) & -\epsilon I \end{bmatrix} < 0,$$
$$(4.5)$$

where $S \in \mathbb{R}^{n \times (n-r)}$ is any matrix with full column rank and satisfies $ES = 0$.

Proof. (*Sufficiency*) Suppose that there exist matrices $P > 0$, Q and a scalar $\epsilon > 0$ such that the LMI in (4.5) holds. Then, by Schur complement, it follows that

$$\left(PE^T + SQ\right)^T A^T + A\left(PE^T + SQ\right) + \epsilon MM^T$$
$$+ \epsilon^{-1}\left(PE^T + SQ\right)^T N_1^T N_1\left(PE^T + SQ\right) < 0. \qquad (4.6)$$

Now, noting (4.2), (4.3) and using Lemma 2.7, we have

$$\left(PE^T + SQ\right)^T \Delta A^T + \Delta A\left(PE^T + SQ\right)$$
$$= \left(PE^T + SQ\right)^T N_1^T F(\sigma)^T M^T + MF(\sigma)N_1\left(PE^T + SQ\right)$$
$$\leq \epsilon MM^T + \epsilon^{-1}\left(PE^T + SQ\right)^T N_1^T N_1\left(PE^T + SQ\right).$$

This together with (4.6) gives

$$\left(PE^T + SQ\right)^T (A + \Delta A)^T + (A + \Delta A)\left(PE^T + SQ\right) < 0.$$

Therefore, with this and Definition 4.2, it is easy to see that the uncertain singular system in (4.1) is generalized quadratically stable.

(*Necessity*) Assume that uncertain singular system (4.1) is generalized quadratically stable. Then, it follows from Definition 4.2 that there exists matrices $P > 0$ and Q such that (4.4) holds for all $F(\sigma)$ satisfying (4.2) and (4.3). Then, it is easy to see that for all $x \in \mathbb{R}^n$ and $x \neq 0$,

$$2x^T \left(PE^T + SQ\right)^T (A + \Delta A)^T x < 0. \tag{4.7}$$

Let

$$Z = \left(PE^T + SQ\right)^T A^T + A \left(PE^T + SQ\right).$$

Then, (4.7) can be rewritten as

$$x^T Z x < -2x^T \left(PE^T + SQ\right)^T N_1^T F(\sigma)^T M^T x, \tag{4.8}$$

for all allowable $F(\sigma)$ and $x \in \mathbb{R}^n$ and $x \neq 0$. Therefore,

$$x^T Z x < -2 \max \left\{ x^T \left(PE^T + SQ\right)^T N_1^T F(\sigma)^T M^T x : F(\sigma)^T F(\sigma) \leq I \right\} \leq 0,$$

which implies

$$\left(x^T Z x\right)^2 > 4 \max \left\{ \left[x^T \left(PE^T + SQ\right)^T N_1^T F(\sigma)^T M^T x \right]^2 : F(\sigma)^T F(\sigma) \leq I \right\}.$$

Applying Lemma 4.2, we have

$$\left(x^T Z x\right)^2 > 4x^T \left(PE^T + SQ\right)^T N_1^T N_1 \left(PE^T + SQ\right) x x^T M M^T x,$$

Thus, by Lemma 4.3, we have that there exists a scalar $\epsilon > 0$ such that

$$\epsilon^2 M M^T + \epsilon Z + \left(PE^T + SQ\right)^T N_1^T N_1 \left(PE^T + SQ\right) < 0.$$

That is,

$$Z + \epsilon M M^T + \epsilon^{-1} \left(PE^T + SQ\right)^T N_1^T N_1 \left(PE^T + SQ\right) < 0,$$

which, by Schur complement, implies that the LMI in (4.5) is satisfied. $\qquad\square$

Remark 4.2. Theorem 4.1 provides a necessary and sufficient condition for the uncertain continuous singular system in (4.1) to be generalized quadratically stable. It is noted that the condition in (4.5) is a strict LMI, which can be checked numerically very efficiently by using recently developed interior-point methods [15]. $\qquad\triangleleft$

In the case when $E = I$, that is, the uncertain singular system in (4.1) reduces to the following uncertain state-space system

$$\dot{x}(t) = (A + \Delta A)\, x(t), \tag{4.9}$$

by Theorem 4.1, it is easy to have the following result.

Corollary 4.1. *The uncertain continuous state-space system in (4.9) is quadratically stable if and only if there exist a matrix $P > 0$ and a scalar $\epsilon > 0$ such that the following LMI holds:*

$$\begin{bmatrix} PA^T + AP + \epsilon MM^T & PN_1^T \\ N_1 P & -\epsilon I \end{bmatrix} < 0. \tag{4.10}$$

Now, we consider a continuous singular system with convex polytopic uncertainties described by

$$E\dot{x}(t) = A(\alpha)x(t), \tag{4.11}$$

where the matrix $A(\alpha) \in \mathbb{R}^{n \times n}$ is not precisely known, which is assumed to belong to a convex bounded (polytopic type) uncertain domain $\mathcal{A}$ defined by

$$\mathcal{A} = \left\{ A(\alpha) \mid A(\alpha) = \sum_{i=1}^{N} \alpha_i A_i; \ \ \sum_{i=1}^{N} \alpha_i = 1; \ \ \alpha_i \geq 0 \right\}. \tag{4.12}$$

For the uncertain singular system in (4.11), we have the following robust stability result.

Theorem 4.2. *The uncertain singular system in (4.11) is robustly stable in uncertainty domain (4.12) if there exist matrices $P > 0$ and Q such that the following LMI holds:*

$$\left(PE^T + SQ\right)^T A_i^T + A_i \left(PE^T + SQ\right) < 0 \tag{4.13}$$

where $S \in \mathbb{R}^{n \times (n-r)}$ is any matrix with full column rank and satisfies $ES = 0$.

Proof. Under the condition of the theorem, it is easy to see that

$$\sum_{i=1}^{N} \alpha_i \left[\left(PE^T + SQ\right)^T A_i^T + A_i \left(PE^T + SQ\right)\right] < 0.$$

That is,

$$\left(PE^T + SQ\right)^T A(\alpha)^T + A(\alpha)\left(PE^T + SQ\right) < 0.$$

With this and Corollary 2.2, we have that the uncertain singular system (4.11) is robustly stable in uncertainty domain (4.12). □

Theorem 4.2 provides a sufficient condition guaranteeing that the uncertain singular system (4.11) is robustly stable. This, however, is conservative since a fixed Lyapunov matrix is used. One way to reduce such a conservatism is to use a parameter-dependent Lyapunov matrix. To show this, we need the following lemma.

Lemma 4.4. *There exist matrices $P > 0$ and Q satisfying*

$$\left(PE^T + SQ\right)^T A^T + A\left(PE^T + SQ\right) < 0, \tag{4.14}$$

if and only if there exist matrices $P > 0$, Q, F and G satisfying

$$\begin{bmatrix} AF^T + FA^T & \left(PE^T + SQ\right)^T + AG - F \\ PE^T + SQ + G^T A^T - F^T & -G - G^T \end{bmatrix} < 0, \tag{4.15}$$

Proof. (*Sufficiency*) Suppose that there exist matrices $P > 0$, Q, F and G such that the LMI in (4.15) holds. Then, pre- and post-multiplying (4.15) by

$$\begin{bmatrix} I & A \end{bmatrix}$$

and its transpose, we obtain (4.14).

(*Necessity*) Suppose that there exist matrices $P > 0$ and Q satisfying the LMI in (4.14). Then, it is easy to see that there exists a scalar $\theta > 0$ such that

$$\begin{bmatrix} \left(PE^T + SQ\right)^T A^T + A\left(PE^T + SQ\right) - \theta AA^T & \theta A \\ \theta A^T & -\theta I \end{bmatrix} < 0. \tag{4.16}$$

Now, choose

$$F = \left(PE^T + SQ\right)^T - \frac{1}{2}\theta A, \quad G = \frac{1}{2}\theta I. \tag{4.17}$$

Then, by (4.16), it can be verified that (4.15) holds for F and G given in (4.17). This completes the proof. □

Now, we are in a position to present a robust stability result via a parameter-dependent Lyapunov matrix.

Theorem 4.3. *The uncertain singular system in (4.11) is robustly stable in uncertainty domain (4.12) if there exist matrices $P_i > 0$, Q_i, $i = 1, \ldots, N$, F and G satisfying*

$$\begin{bmatrix} A_iF^T + FA_i^T & \left(P_iE^T + SQ_i\right)^T + A_iG - F \\ P_iE^T + SQ_i + G^TA_i^T - F^T & -G - G^T \end{bmatrix} < 0, \quad (4.18)$$

for $i = 1, \ldots, N$, where $S \in \mathbb{R}^{n \times (n-r)}$ is any matrix with full column rank and satisfies $ES = 0$.

Proof. By (4.12) and (4.18), it is easy to see that

$$\sum_{i=1}^{N} \alpha_i \begin{bmatrix} A_iF^T + FA_i^T & \left(P_iE^T + SQ_i\right)^T + A_iG - F \\ P_iE^T + SQ_i + G^TA_i^T - F^T & -G - G^T \end{bmatrix} < 0.$$

That is,

$$\begin{bmatrix} A(\alpha)F^T + FA(\alpha)^T & \\ P(\alpha)E^T + SQ(\alpha) + G^TA(\alpha)^T - F^T & \\ & \left(P(\alpha)E^T + SQ(\alpha)\right)^T + A(\alpha)G - F \\ & -G - G^T \end{bmatrix} < 0, \quad (4.19)$$

where

$$P(\alpha) = \sum_{i=1}^{N} \alpha_i P_i, \quad Q(\alpha) = \sum_{i=1}^{N} \alpha_i Q_i.$$

Now, applying Lemma 4.4 to (4.19) we have

$$[P(\alpha)E + SQ(\alpha)]^T A(\alpha) + A(\alpha)^T [P(\alpha)E + SQ(\alpha)] < 0.$$

With this and Corollary 2.2, it is easy to see that the uncertain singular system (4.11) is robustly stable in uncertainty domain (4.12). □

To show the reduced conservatism of Theorem 4.3, we provide the following example.

Example 4.1. Consider an uncertain continuous singular system in (4.13) with

$$E = \begin{bmatrix} 0 & -1 & 1 \\ 0 & -2 & 2 \\ 0 & -1 & 1 \end{bmatrix}.$$

We assume that there are three vertices in (4.13), which are given as

$$A_1 = \begin{bmatrix} -1 & 1.9 & -2.9 \\ -1 & -1.2 & -3.8 \\ 1 & 7.9 & 1.1 \end{bmatrix}, \quad A_2 = \begin{bmatrix} -1 & -7.4 & -2.6 \\ -1 & 1.2 & -3.2 \\ 1 & -7.4 & 1.4 \end{bmatrix},$$

$$A_3 = \begin{bmatrix} -1 & 1.1 & -2.1 \\ -1 & -2.8 & -2.2 \\ 1 & 7.1 & 1.9 \end{bmatrix}.$$

Now, choose

$$S = \begin{bmatrix} -4 & 1 \\ 1 & 0 \\ 1 & 0 \end{bmatrix},$$

which is with full column rank and satisfies $ES = 0$. Then, it can be verified that the robust stability condition in Theorem 4.2 is not feasible; therefore, it fails to conclude whether this uncertain system is robustly stable or not by the criteria in Theorem 4.2. However, by using the Matlab LMI Control Toolbox, it is found that the LMIs given in (4.18) are feasible. Therefore, we have that the uncertain singular system in this example is robustly stable. This shows that the robust stability in Theorem 4.3 is less conservative than that in Theorem 4.2.   ◇

4.2.2 Discrete Systems

In this section, we investigate the robust stability problem for uncertain discrete singular systems. Firstly, we consider an uncertain linear discrete-time singular system with time-invariant norm-bounded parameter uncertainties, which is described by

$$Ex(k+1) = (A + \Delta A)x(k), \tag{4.20}$$

where $x(k) \in \mathbb{R}^n$ is the state. The matrix $E \in \mathbb{R}^{n \times n}$ may be singular; we shall assume that $\mathrm{rank}(E) = r \leq n$. A is a known real constant matrix with appropriate dimensions. ΔA is an unknown time-invariant matrices representing norm-bounded parameter uncertainties, which satisfies (4.2) and (4.3). Similar to the continuous case, we introduce the following definition.

Definition 4.3. *The uncertain discrete singular system in (4.20) is said to be robustly stable if it is regular, causal and stable for all allowable uncertainties.*

Considering that the concept of quadratic stability for uncertain discrete state-space systems has played an important role in dealing with the robust stability problem, we introduce the following definition.

Definition 4.4. *The uncertain discrete singular system in (4.20) is said to be generalized quadratically stable if there exist matrices $P > 0$ and Q such that*

$$(A + \Delta A)^T P (A + \Delta A) - E^T P E$$
$$+ Q S^T (A + \Delta A) + (A + \Delta A)^T S Q^T < 0, \tag{4.21}$$

where $S \in \mathbb{R}^{n \times (n-r)}$ is any matrix with full column rank and satisfies $E^T S = 0$.

It is easy to show that generalized quadratic stability implies robust stability of the uncertain system in (4.20) for all allowable parameter uncertainties. Taking this into account, we provide a necessary and sufficient condition for generalized quadratic stability in the following theorem.

Theorem 4.4. *The uncertain discrete singular system in (4.20) is generalized quadratically stable if and only if there exist a scalar $\epsilon > 0$, matrices $P > 0$ and Q such that the following LMI holds:*

$$\begin{bmatrix} QS^T A + A^T SQ^T - E^T PE + \epsilon N_1^T N_1 & A^T P & QS^T M \\ PA & -P & PM \\ M^T SQ^T & M^T P & -\epsilon I \end{bmatrix} < 0, \qquad (4.22)$$

where $S \in \mathbb{R}^{n \times (n-r)}$ is any matrix with full column rank and satisfies $E^T S = 0$.

Proof. (*Sufficiency*) Suppose that there exist a scalar $\epsilon > 0$, matrices $P > 0$ and Q such that the LMI in (4.22) holds. Then, by Schur complement, we have

$$\begin{bmatrix} QS^T A + A^T SQ^T - E^T PE & A^T P \\ PA & -P \end{bmatrix}$$
$$+\epsilon^{-1} \begin{bmatrix} QS^T M \\ PM \end{bmatrix} \begin{bmatrix} QS^T M \\ PM \end{bmatrix}^T + \epsilon \begin{bmatrix} N_1^T \\ 0 \end{bmatrix} \begin{bmatrix} N_1^T \\ 0 \end{bmatrix}^T < 0. \qquad (4.23)$$

Now, using Lemma 2.7, we have

$$\begin{bmatrix} QS^T \Delta A + \Delta A^T SQ^T & \Delta A^T P \\ P\Delta A & 0 \end{bmatrix}$$
$$= \begin{bmatrix} QS^T M \\ PM \end{bmatrix} F(\sigma) \begin{bmatrix} N_1 & 0 \end{bmatrix} + \begin{bmatrix} N_1^T \\ 0 \end{bmatrix} F(\sigma)^T \begin{bmatrix} M^T SQ^T & M^T P \end{bmatrix}$$
$$\leq \epsilon^{-1} \begin{bmatrix} QS^T M \\ PM \end{bmatrix} \begin{bmatrix} QS^T M \\ PM \end{bmatrix}^T + \epsilon \begin{bmatrix} N_1^T \\ 0 \end{bmatrix} \begin{bmatrix} N_1^T \\ 0 \end{bmatrix}^T.$$

Hence,

$$\begin{bmatrix} QS^T (A + \Delta A) + (A + \Delta A)^T SQ^T - E^T PE & (A + \Delta A)^T P \\ P(A + \Delta A) & -P \end{bmatrix}$$
$$\leq \begin{bmatrix} QS^T A + A^T SQ^T - E^T PE & A^T P \\ PA & -P \end{bmatrix}$$
$$+\epsilon^{-1} \begin{bmatrix} QS^T M \\ PM \end{bmatrix} \begin{bmatrix} QS^T M \\ PM \end{bmatrix}^T + \epsilon \begin{bmatrix} N_1^T \\ 0 \end{bmatrix} \begin{bmatrix} N_1^T \\ 0 \end{bmatrix}^T.$$

This together with (4.23) gives

$$\begin{bmatrix} QS^T (A + \Delta A) + (A + \Delta A)^T SQ^T - E^T PE & (A + \Delta A)^T P \\ P(A + \Delta A) & -P \end{bmatrix} < 0,$$

which, by Schur complement again, implies

$$(A + \Delta A)^T P (A + \Delta A) - E^T PE$$
$$+ QS^T (A + \Delta A) + (A + \Delta A)^T SQ^T < 0.$$

Then, by Definition 4.4, we have that the uncertain discrete-time singular system (Σ) is generalized quadratically stable.

(*Necessity*) Assume that the uncertain discrete singular system (4.20) is generalized quadratically stable. Then, it follows from Definition 4.4 that there exist matrices $P > 0$ and Q such that (4.21) holds. Thus, for all $F(\sigma)$ satisfying (4.2) and (4.3), the following inequality holds:

$$\begin{bmatrix} QS^T (A + \Delta A) + (A + \Delta A)^T SQ^T - E^T PE & (A + \Delta A)^T P \\ P(A + \Delta A) & -P \end{bmatrix} < 0.$$

That is,

$$\begin{bmatrix} QS^T A + A^T SQ^T - E^T PE & A^T P \\ PA & -P \end{bmatrix}$$
$$+ \begin{bmatrix} QS^T M \\ PM \end{bmatrix} F(\sigma) \begin{bmatrix} N_1 & 0 \end{bmatrix}$$
$$+ \begin{bmatrix} N_1^T \\ 0 \end{bmatrix} F(\sigma)^T \begin{bmatrix} M^T SQ^T & M^T P \end{bmatrix} < 0, \qquad (4.24)$$

is satisfied for all $F(\sigma)$ satisfying (4.2) and (4.3). Noting (4.24) and following a similar line as in the proof of Theorem 4.1, we can deduce that there exists a scalar $\epsilon > 0$ such that

$$\begin{bmatrix} QS^T A + A^T SQ^T - E^T PE & A^T P \\ PA & -P \end{bmatrix}$$
$$+ \epsilon^{-1} \begin{bmatrix} QS^T M \\ PM \end{bmatrix} \begin{bmatrix} QS^T M \\ PM \end{bmatrix}^T + \epsilon \begin{bmatrix} N_1^T \\ 0 \end{bmatrix} \begin{bmatrix} N_1^T \\ 0 \end{bmatrix}^T < 0,$$

which, by Schur complement, gives that the LMI in (4.22) holds. This completes the proof. $\qquad \square$

In the case when $E = I$, that is, the uncertain singular system in (4.20) reduces to the following uncertain state-space system

$$x(k + 1) = (A + \Delta A)x(k), \qquad (4.25)$$

by Theorem 4.4, it is easy to have the following result.

Corollary 4.2. *The uncertain discrete state-space system in (4.24) is quadratically stable if and only if there exist a matrix $P > 0$ and a scalar $\epsilon > 0$ such that the following LMI holds:*

$$\begin{bmatrix} \epsilon N_1^T N_1 - P & A^T P & 0 \\ PA & -P & PM \\ 0 & M^T P & -\epsilon I \end{bmatrix} < 0.$$

Now, we consider a discrete singular system with convex polytopic uncertainties described by

$$Ex(k+1) = A(\alpha)x(k), \tag{4.26}$$

where the matrix $A(\alpha) \in \mathbb{R}^{n \times n}$ is not precisely known, but is assumed to belong to a convex bounded (polytopic type) uncertain domain $\mathcal{A}$ defined as in (4.12). Then, for the uncertain discrete singular system in (4.26), we have the following robust stability result.

Theorem 4.5. *The uncertain discrete singular system in (4.26) is robustly stable in uncertainty domain (4.12) if there exist matrices $P > 0$ and Q such that the following LMI holds:*

$$\begin{bmatrix} QS^T A_i + A_i^T SQ^T - E^T PE & A_i^T P \\ PA_i & -P \end{bmatrix} < 0, \tag{4.27}$$

where $S \in \mathbb{R}^{n \times (n-r)}$ is any matrix with full column rank and satisfies $E^T S = 0$.

Proof. By (4.12) and (4.27), it is easy to see that

$$\sum_{i=1}^{N} \alpha_i \begin{bmatrix} QS^T A_i + A_i^T SQ^T - E^T PE & A_i^T P \\ PA_i & -P \end{bmatrix} < 0.$$

That is,

$$\begin{bmatrix} QS^T A(\alpha) + A(\alpha)^T SQ^T - E^T PE & A(\alpha)^T P \\ PA(\alpha) & -P \end{bmatrix} < 0.$$

Applying Schur complement to this LMI provides

$$A(\alpha)^T PA(\alpha) - E^T PE + QS^T A(\alpha) + A(\alpha)^T SQ^T < 0,$$

With this and Theorem 2.4, we have that the uncertain singular system (4.26) is robustly stable in uncertainty domain (4.12). □

Theorem 4.5 is obtained by using a fixed Lyapunov matrix, which may be conservative. One way to reduce such a conservatism is to use a parameter-dependent Lyapunov matrix. To show this, we need the following lemma.

Lemma 4.5. *There exist matrices $P > 0$ and Q satisfying*

$$A^T P A - E^T P E + Q S^T A + A^T S Q^T < 0, \tag{4.28}$$

if and only if there exist matrices $P > 0$, Q, F and G satisfying

$$\begin{bmatrix} A^T F^T + F A - E^T P E & Q S^T + A^T G - F \\ S Q^T + G^T A - F^T & P - G - G^T \end{bmatrix} < 0. \tag{4.29}$$

Proof. (*Sufficiency*) Suppose that there exist matrices $P > 0$, Q, F and G such that the LMI in (4.29) holds. Then, pre- and post-multiplying (4.29) by

$$\begin{bmatrix} I & A^T \end{bmatrix}$$

and its transpose, we obtain (4.28) immediately.

(*Necessity*) Suppose that there exist matrices $P > 0$ and Q satisfying the LMI in (4.28). Then, by Schur complement, it follows from (4.28) that

$$\begin{bmatrix} Q S^T A + A^T S Q^T - E^T P E & A^T P \\ P A & -P \end{bmatrix} < 0. \tag{4.30}$$

Now, choose

$$G = P, \quad F = Q S^T - A^T G^T + A^T P. \tag{4.31}$$

Then, it is easy to see that (4.29) holds for F and G given in (4.31). This completes the proof. $\qquad\square$

Now, we are in a position to present a robust stability result via a parameter-dependent Lyapunov matrix.

Theorem 4.6. *The uncertain singular system in (4.26) is robustly stable in uncertainty domain (4.12) if there exist matrices $P_i > 0$, Q_i, $i = 1, \ldots, N$, F and G satisfying*

$$\begin{bmatrix} A_i^T F^T + F A_i - E^T P_i E & Q_i S^T + A_i^T G - F \\ S Q_i^T + G^T A_i - F^T & P_i - G - G^T \end{bmatrix} < 0, \tag{4.32}$$

for $i = 1, \ldots, N$, where $S \in \mathbb{R}^{n \times (n-r)}$ is any matrix with full column rank and satisfies $E^T S = 0$.

Proof. It follows from (4.12) and (4.32) that

$$\sum_{i=1}^{N} \alpha_i \begin{bmatrix} A_i^T F^T + F A_i - E^T P_i E & Q_i S^T + A_i^T G - F \\ S Q_i^T + G^T A_i - F^T & P_i - G - G^T \end{bmatrix} < 0,$$

which can be rewritten as

$$\begin{bmatrix} A(\alpha)^T F^T + F A(\alpha) - E^T P(\alpha) E & Q(\alpha) S^T + A(\alpha)^T G - F \\ S Q(\alpha)^T + G^T A(\alpha) - F^T & P(\alpha) - G - G^T \end{bmatrix} < 0 \qquad (4.33)$$

where

$$P(\alpha) = \sum_{i=1}^{N} \alpha_i P_i, \quad Q(\alpha) = \sum_{i=1}^{N} \alpha_i Q_i.$$

Then, applying Lemma 4.5 to (4.33), we have

$$A(\alpha)^T P(\alpha) A(\alpha) - E^T P(\alpha) E + Q(\alpha) S^T A(\alpha) + A(\alpha)^T S Q(\alpha)^T < 0.$$

Finally, by Theorem 2.4, it is easy to see that the uncertain singular system (4.11) is robustly stable in uncertainty domain (4.12). □

4.3 Robust Stabilization

In this section, we study the problem of robust stabilization for uncertain singular systems. Via the notion of generalized quadratic stabilizability, the problem is solved for both continuous and discrete singular systems.

4.3.1 Continuous Systems

Consider an uncertain linear continuous singular system described by

$$E\dot{x}(t) = (A + \Delta A)\, x(t) + (B + \Delta B) u(t), \qquad (4.34)$$

where $x(t) \in \mathbb{R}^n$ is the state; $u(t) \in \mathbb{R}^m$ is the control input. The matrix $E \in \mathbb{R}^{n \times n}$ may be singular; we shall assume that $\text{rank}(E) = r \leq n$. A and B are known real constant matrices with appropriate dimensions. ΔA and ΔB are unknown time-invariant matrices representing norm-bounded parameter uncertainties, which are assumed to be of the form:

$$\begin{bmatrix} \Delta A & \Delta B \end{bmatrix} = M F(\sigma) \begin{bmatrix} N_1 & N_2 \end{bmatrix} \qquad (4.35)$$

where M, N_1 and N_2 are known real constant matrices with appropriate dimensions. The uncertain matrix $F(\sigma)$ satisfies (4.3).

To consider the robust stabilization problem of system (4.34), we first introduce the following definition.

Definition 4.5. *The uncertain continuous singular system in (4.34) is said to be robustly stabilizable if there exists a linear state feedback control law*

$$u(t) = Kx(t), \quad K \in \mathbb{R}^{m \times n}, \tag{4.36}$$

such that the closed-loop system is robustly stable in the sense of Definition 4.1. In this case, (4.36) is said to be a robust state feedback control law for the uncertain system in (4.34).

Similar to the study of stability robustness, we introduce the following definition.

Definition 4.6. *The uncertain continuous singular system in (4.34) is said to be generalized quadratically stabilizable if there exists a linear state feedback control law (4.36) such that the closed-loop system is quadratically stable in the sense of Definition 4.2.*

Similar to the derivation of Lemma 4.1, it can be shown that generalized quadratic stabilizability implies robust stabilizability. Taking this into account, our attention will be focused on the development of a condition for generalized quadratic stabilization for the uncertain continuous singular system in (4.34); such a condition is given in the following theorem.

Theorem 4.7. *The uncertain continuous singular system in (4.34) is generalized quadratically stabilizable if and only if there exist matrices $P > 0$, Q, Y and a scalar $\epsilon > 0$ such that*

$$\begin{bmatrix} \Gamma(P,Q) + \epsilon M M^T & \Omega(P,Q)^T N_1^T + Y^T N_2^T \\ N_1 \Omega(P,Q) + N_2 Y & -\epsilon I \end{bmatrix} < 0, \tag{4.37}$$

where

$$\Gamma(P,Q) = \Omega(P,Q)^T A^T + A\Omega(P,Q) + BY + Y^T B^T, \tag{4.38}$$

$$\Omega(P,Q) = PE^T + SQ, \tag{4.39}$$

and $S \in \mathbb{R}^{n \times (n-r)}$ is any matrix with full column rank and satisfies $ES = 0$. In this case, we can assume that the matrix $\Omega(P,Q)$ is nonsingular (if this is not the case, then we can choose some $\theta \in (0,1)$ such that $\hat{\Omega}(P,Q) = \Omega(P,Q) + \theta \tilde{P}$ is nonsingular and satisfies (3.4), in which $\tilde{P}$ is any nonsingular matrix satisfying $E\tilde{P} = \tilde{P}^T E^T \geq 0$), then a desired robustly stabilizing state feedback controller can be chosen as

$$u(t) = Y\Omega(P,Q)^{-1} x(t). \tag{4.40}$$

Proof. (*Sufficiency*) Suppose that there exist matrices $P > 0$, Q, Y and a scalar $\epsilon > 0$ such that (4.37) holds. Then, by Schur complement, it follows from (4.37) that

$$\Gamma(P,Q) + \epsilon M M^T$$
$$+ \epsilon^{-1} \left[\Omega(P,Q)^T N_1^T + Y^T N_2^T \right] [N_1 \Omega(P,Q) + N_2 Y] < 0. \qquad (4.41)$$

Now, applying the state feedback controller in (4.40) to (4.34), we have the closed-loop system as

$$E\dot{x}(t) = (A_c + \Delta A_c)\, x(t), \qquad (4.42)$$

where

$$A_c = A + BY\Omega(P,Q)^{-1},$$
$$\Delta A_c = MF(\sigma)\left(N_1 + N_2 Y\Omega(P,Q)^{-1}\right).$$

By (4.41) and Lemma 2.7, it can be shown that

$$\left(PE^T + SQ\right)^T (A_c + \Delta A_c)^T + (A_c + \Delta A_c)\left(PE^T + SQ\right) < 0.$$

Then, by Definition 4.6, it is easy to see that the uncertain continuous singular system (4.34) is generalized quadratically stabilizable.

(*Necessity*) Suppose that the uncertain continuous singular system (4.34) is generalized quadratically stabilizable. Then, by Definition 4.6, we have that there exists a linear state feedback control law (4.36) such that the closed-loop system is quadratically stable, which implies that there exist matrices $P > 0$ and Q such that

$$\left(PE^T + SQ\right)^T (A_K + \Delta A_K)^T + (A_K + \Delta A_K)\left(PE^T + SQ\right) < 0, \qquad (4.43)$$

where

$$A_K = A + BK, \qquad \Delta A_K = \Delta A + \Delta BK.$$

Then, along a similar line as in the proof of the necessity part of Theorem 4.1, we can deduce that there exists a scalar $\epsilon > 0$ such that

$$\left[\begin{array}{cc} \left(PE^T + SQ\right)^T A_K^T + A_K\left(PE^T + SQ\right) + \epsilon M M^T & \left(PE^T + SQ\right)^T (N_1 + N_2 K)^T \\ (N_1 + N_2 K)\left(PE^T + SQ\right) & -\epsilon I \end{array} \right] < 0. \qquad (4.44)$$

Now, set

$$Y = K\left(PE^T + SQ\right).$$

Then, it is to see that (4.44) provides (4.37). This completes the proof. $\qquad\square$

Now, we provide an example to show the applicability of Theorem 4.7.

Example 4.2. Consider an uncertain continuous singular system in (4.34) with

$$E = \begin{bmatrix} 2 & -3 & 1 \\ 4 & 1 & 2 \\ 2 & -5 & 1 \end{bmatrix}, \quad A = \begin{bmatrix} 2 & 7 & 1 \\ 4 & 0 & 2 \\ 2 & 11 & 1 \end{bmatrix},$$

$$B = \begin{bmatrix} 2.1 & -0.3 \\ 2.3 & 0.2 \\ 1.5 & 1.3 \end{bmatrix}, \quad M = \begin{bmatrix} -0.2 & 0.4 \\ 0.3 & -1 \\ -0.4 & 1.6 \end{bmatrix},$$

$$N_1 = \begin{bmatrix} 0.2 & 0.8 & 1.1 \\ -0.1 & 0 & -0.3 \end{bmatrix}, \quad N_2 = \begin{bmatrix} 0.3 & -0.5 \\ 0 & 0.6 \end{bmatrix}$$

Then, it can be verified that the nominal system is not regular, unstable and impulsive. Note that $\text{rank}(E) = 2$, thus we can choose

$$S = \begin{bmatrix} 1 \\ 0 \\ -2 \end{bmatrix},$$

which is with full column rank and satisfies $ES = 0$. Now, it is found that the LMI in (4.37) is feasible and a set of solutions can be obtained as follows:

$$P = \begin{bmatrix} 0.6188 & 0.0466 & -0.3282 \\ 0.0466 & 0.1160 & 0.0233 \\ -0.3282 & 0.0233 & 1.1110 \end{bmatrix},$$

$$Y = \begin{bmatrix} -0.2968 & -4.7198 & -0.3678 \\ 1.6909 & -0.7614 & 0.7572 \end{bmatrix},$$

$$Q = \begin{bmatrix} -0.3520 & 0.4111 & -0.1994 \end{bmatrix},$$

$$\epsilon = 1.7353.$$

Therefore, by Theorem 4.7, a desired stabilizing state feedback controller can be constructed as

$$u(t) = \begin{bmatrix} -1.9807 & -0.7113 & 0.3357 \\ -0.5680 & 0.9025 & 1.9626 \end{bmatrix} x(t).$$

$\diamond$

In the case when $E = I$, that is, the uncertain singular system in (4.34) reduces to the following uncertain state-space system

$$\dot{x}(t) = (A + \Delta A)\, x(t) + (B + \Delta B)u(t), \tag{4.45}$$

by Theorem 4.7, it is easy to have the following result.

Corollary 4.3. *The uncertain continuous state-space system in (4.45) is quadratically stabilizable if and only if there exist matrices $P > 0$, Y and a scalar $\epsilon > 0$ such that*

$$\begin{bmatrix} PA^T + AP + BY + Y^TB^T + \epsilon MM^T & PN_1^T + Y^TN_2^T \\ N_1P + N_2Y & -\epsilon I \end{bmatrix} < 0.$$

In this case, a desired robustly stabilizing state feedback controller may be chosen as

$$u(t) = YP^{-1}x(t).$$

4.3.2 Discrete Systems

Now, we consider the robust stabilization problem for uncertain discrete singular systems via state feedback controllers. The class of uncertain discrete singular system to be considered is described by

$$Ex(k+1) = (A + \Delta A)x(k) + (B + \Delta B)u(k), \tag{4.46}$$

where $x(k) \in \mathbb{R}^n$ is the state; $u(k) \in \mathbb{R}^m$ is the control input. The matrix $E \in \mathbb{R}^{n \times n}$ may be singular; we shall assume that $\text{rank}(E) = r \leq n$. A and B are known real constant matrices with appropriate dimensions. ΔA and ΔB are time-invariant matrices representing norm-bounded parameter uncertainties, and are assumed to be of the form in (4.35).

To consider the robust stabilization problem of system (4.46), we introduce the following definitions.

Definition 4.7. *The uncertain singular system in (4.46) is said to be robustly stabilizable if there exists a linear state feedback control law*

$$u(k) = Kx(k), \quad K \in \mathbb{R}^{m \times n}, \tag{4.47}$$

such that the closed-loop system is robustly stable in the sense of Definition 4.3. In this case, (4.47) is said to be a robust state feedback control law for the uncertain singular system in (4.46).

Definition 4.8. *The uncertain discrete-time singular system in (4.46) is said to be generalized quadratically stabilizable if there exists a linear state feedback control law (4.47) such that the closed-loop system is quadratically stable in the sense of Definition 4.4.*

It can be shown that generalized quadratic stabilizability implies robust stabilizability. In order to obtain the condition for generalized quadratic stabilizability for the uncertain discrete-time singular system in (4.46), we first introduce the following lemma.

Lemma 4.6. [173] *Given any matrices $\mathcal{X}$, $\mathcal{Y}$ and $\mathcal{Z}$ with appropriate dimensions such that $\mathcal{Y} > 0$. Then we have*

$$\mathcal{X}^T \mathcal{Z} + \mathcal{Z}^T \mathcal{X} + \mathcal{X}^T \mathcal{Y} \mathcal{X} \geq -\mathcal{Z}^T \mathcal{Y}^{-1} \mathcal{Z}.$$

Now, we present the generalized quadratic stability result in the following theorem.

Theorem 4.8. *The uncertain discrete-time singular system in (4.46) is generalized quadratically stabilizable if and only if there exist scalars $\epsilon > 0$, $\delta > 0$, matrices $P > 0$ and Q such that*

$$\Gamma \triangleq P^{-1} - \epsilon^{-1} M M^T > 0, \tag{4.48}$$

and

$$QS^T A + A^T SQ^T - E^T PE + \epsilon N_1^T N_1 + \epsilon^{-1} QS^T MM^T SQ^T$$
$$+ \left(A^T + \epsilon^{-1} QS^T MM^T\right) \Gamma^{-1} \left(A^T + \epsilon^{-1} QS^T MM^T\right)^T$$
$$- \Psi \Phi^{-1} \Psi^T < 0, \tag{4.49}$$

where the matrix $S \in \mathbb{R}^{n \times (n-r)}$ is any matrix with full column rank and satisfies $E^T S = 0$, and

$$\Phi \triangleq B^T \Gamma^{-1} B + \epsilon N_2^T N_2 + \delta I,$$
$$\Psi \triangleq QS^T B + \left(A^T + \epsilon^{-1} QS^T MM^T\right) \Gamma^{-1} B + \epsilon N_1^T N_2.$$

In this case, a desired robustly stabilizing state feedback control law can be chosen by

$$u(k) = -\Phi^{-1} \Psi^T x(k). \tag{4.50}$$

Proof. (Sufficiency) Applying the state feedback controller (4.50) to (4.46) results in the following closed-loop system

$$Ex(k+1) = [A_c + MF(\sigma)N_c] x(k), \tag{4.51}$$

where

$$A_c = A - B\Phi^{-1}\Psi^T, \quad N_c = N_1 - N_2\Phi^{-1}\Psi^T.$$

Note that

$$B^T \Gamma^{-1} B + \epsilon N_2^T N_2 < \Phi.$$

Then, by a routine calculation, it can be verified that

$$
\begin{aligned}
& QS^T A_c + A_c^T SQ^T + \epsilon N_c^T N_c \\
& + \left(A_c^T + \epsilon^{-1} QS^T MM^T\right) \Gamma^{-1} \left(A_c^T + \epsilon^{-1} QS^T MM^T\right)^T \\
& \leq QS^T A_c + A_c^T SQ^T + \epsilon N_1^T N_1 - \epsilon N_1^T N_2 \Phi^{-1} \Psi^T - \epsilon \Psi \Phi^{-1} N_2^T N_1 \\
& + \left(A^T + \epsilon^{-1} QS^T MM^T\right) \Gamma^{-1} \left(A^T + \epsilon^{-1} QS^T MM^T\right)^T \\
& - \left(A^T + \epsilon^{-1} QS^T MM^T\right) \Gamma^{-1} B \Phi^{-1} \Psi^T \\
& - \Psi \Phi^{-1} B^T \Gamma^{-1} \left(A^T + \epsilon^{-1} QS^T MM^T\right)^T + \Psi \Phi^{-1} \Psi^T \\
& = QS^T A + A^T SQ^T + \epsilon N_1^T N_1 - \Psi \Phi^{-1} \Psi^T \\
& + \left(A^T + \epsilon^{-1} QS^T MM^T\right) \Gamma^{-1} \left(A^T + \epsilon^{-1} QS^T MM^T\right)^T.
\end{aligned}
$$

This together with (4.49) gives

$$
\begin{aligned}
& QS^T A_c + A_c^T SQ^T - E^T PE + \epsilon N_c^T N_c + \epsilon^{-1} QS^T MM^T SQ^T \\
& + \left(A_c^T + \epsilon^{-1} QS^T MM^T\right) \Gamma^{-1} \left(A_c^T + \epsilon^{-1} QS^T MM^T\right)^T < 0,
\end{aligned}
$$

which, by Schur complement, implies

$$
\begin{bmatrix}
QS^T A_c + A_c^T SQ^T - E^T PE + \epsilon N_c^T N_c & A_c^T P & QS^T M \\
PA_c & -P & PM \\
M^T SQ^T & M^T P & -\epsilon I
\end{bmatrix} < 0.
$$

Thus, by Theorem 4.4, we have that the closed-loop system (4.51) is generalized quadratically stable which, by Definition 4.8, implies that the uncertain discrete-time singular system in (4.46) is generalized quadratically stabilizable.

(*Necessity*) Assume that the uncertain discrete-time singular system in (4.46) is generalized quadratically stabilizable. Then, by Definition 4.8, it follows that there exist a linear state feedback control law $u(k) = Kx(k)$, $K \in \mathbb{R}^{m \times n}$, matrices $P > 0$ and Q such that

$$
\begin{aligned}
& (A_K + \Delta A_K)^T P (A_K + \Delta A_K) - E^T PE \\
& + QS^T (A_K + \Delta A_K) + (A_K + \Delta A_K)^T SQ^T < 0,
\end{aligned}
\tag{4.52}
$$

is satisfied for all allowable uncertainties ΔA and ΔB, where

$$A_K = A + BK, \quad \Delta A_K = \Delta A + \Delta BK. \tag{4.53}$$

Thus, by Theorem 4.4, we have that there exists a scalar $\epsilon > 0$ such that

$$
\begin{bmatrix}
QS^T A_K + A_K^T SQ^T - E^T PE + \epsilon N_K^T N_K & A_K^T P & QS^T M \\
PA_K & -P & PM \\
M^T SQ^T & M^T P & -\epsilon I
\end{bmatrix} < 0. \tag{4.54}
$$

where A_K is given in (4.53) and

$$N_K = N_1 + N_2 K.$$

By Schur complement, it follows from (4.54) that

$$QS^T A_K + A_K^T SQ^T - E^T PE$$
$$+\epsilon N_K^T N_K + \epsilon^{-1} QS^T MM^T SQ^T$$
$$+ \left(A_K^T + \epsilon^{-1} QS^T MM^T \right) \Gamma^{-1} \left(A_K^T + \epsilon^{-1} QS^T MM^T \right)^T < 0. \quad (4.55)$$

Substituting the expressions of A_K and N_K into (4.55) yields

$$QS^T A + A^T SQ^T - E^T PE + \epsilon N_1^T N_1 + \epsilon^{-1} QS^T MM^T SQ^T$$
$$+ \left(A^T + \epsilon^{-1} QS^T MM^T \right) \Gamma^{-1} \left(A^T + \epsilon^{-1} QS^T MM^T \right)^T$$
$$+ \Psi K + K^T \Psi^T + K^T \left[B^T \Gamma^{-1} B + \epsilon N_2^T N_2 \right] K < 0.$$

From this inequality, it is easy to show that there exists a scalar $\delta > 0$ such that

$$QS^T A + A^T SQ^T - E^T PE + \epsilon N_1^T N_1 + \epsilon^{-1} QS^T MM^T SQ^T$$
$$+ \left(A^T + \epsilon^{-1} QS^T MM^T \right) \Gamma^{-1} \left(A^T + \epsilon^{-1} QS^T MM^T \right)^T$$
$$+ \Psi K + K^T \Psi^T + K^T \left[B^T \Gamma^{-1} B + \epsilon N_2^T N_2 + \delta I \right] K < 0. \quad (4.56)$$

By Lemma 4.6, it follows that

$$\Psi K + K^T \Psi^T + K^T \left[B^T \Gamma^{-1} B + \epsilon N_2^T N_2 + \delta I \right] K \geq -\Psi \Phi^{-1} \Psi^T.$$

This together with (4.56) implies that (4.49) holds, which completes the proof.

$$\square$$

Now, we provide an illustrative example.

Example 4.3. Consider an uncertain discrete-time singular system in (4.46) with parameters as follows:

$$E = \begin{bmatrix} 1 & 0 & 0.5 \\ 2 & 1 & 1 \\ 0 & 0 & 0 \end{bmatrix}, \quad A = \begin{bmatrix} 2.4 & 0.2 & 1.2 \\ 4 & 1.5 & 2 \\ 0 & 0 & 0 \end{bmatrix},$$

$$B = \begin{bmatrix} 0 & 1 & 1 \\ 1 & 0 & 0 \\ 1 & 2 & 1 \end{bmatrix}, \quad M = \begin{bmatrix} 0.1 \\ 0.2 \\ -0.1 \end{bmatrix}$$

$$N_1 = \begin{bmatrix} 0.35 & 0.4 & 0.7 \end{bmatrix}, \quad N_2 = \begin{bmatrix} 0.1 & 0.2 & 0 \end{bmatrix}.$$

It can be verified that the nominal discrete-singular system is not regular, non-causal, and unstable. The purpose is the design of a state feedback control law such that the closed-loop system is robustly stable for all allowable uncertainties. To this end, we choose

$$S = \begin{bmatrix} 0 & 0 & 1 \end{bmatrix}^T.$$

Then, it can be verified that

$$P = \begin{bmatrix} 2 & 0 & 1 \\ 0 & 2 & 1 \\ 1 & 1 & 3 \end{bmatrix}, \quad Q = \begin{bmatrix} 0.7 \\ -4.4 \\ -8.4 \end{bmatrix},$$
$$\delta = 0.01, \quad \epsilon = 1.2,$$

satisfy the matrix inequalities in (4.48) and (4.49). Therefore, by Theorem 4.8, a desired robustly stabilizing state feedback control law can be chosen as

$$u(k) = \begin{bmatrix} -3.5469 & -2.2978 & -3.5197 \\ 4.7353 & 5.0297 & 9.5332 \\ -6.5998 & -5.9272 & -12.0526 \end{bmatrix} x(k).$$

◇

In the case when $E = I$, that is, the uncertain singular system (4.46) reduces to the following uncertain state-space system

$$x(k+1) = (A + \Delta A)x(k) + (B + \Delta B)u(k), \tag{4.57}$$

from the proof of Theorem 4.8, it is easy to have the following result.

Corollary 4.4. *The uncertain discrete state-space system in (4.57) is quadratically stabilizable if and only if there exist a scalar $\epsilon > 0$, matrices $P > 0$ and Y such that*

$$\begin{bmatrix} -P & PA^T + Y^T B^T & PN_1^T + Y^T N_2^T \\ AP + BY & \epsilon M M^T - P & 0 \\ N_1 P + N_2 Y & 0 & -\epsilon I \end{bmatrix} < 0.$$

In this case, a desired robustly stabilizing state feedback control law can be chosen by

$$u(k) = Y P^{-1} x(k).$$

4.4 Conclusion

This chapter has studied the problems of robust stability analysis and robust stabilization for both continuous and discrete singular systems. The notions of generalized quadratic stability and generalized quadratic stabilizability have been proposed. It has been shown that generalized quadratic stability and generalized quadratic stabilizability imply robust stability and robust stabilizability, respectively. In both the contexts of continuous and discrete singular systems, necessary and sufficient conditions for generalized quadratic stability and generalized quadratic stabilizability have been obtained. Furthermore, explicit expressions of desired robustly stabilizing state feedback controllers have been given. Part of the results presented in this chapter have appeared in [179].

5

H_∞ Control

5.1 Introduction

The problem of H_∞ control has been a topic of recurring interest for two decades. The so-called bounded real lemma has played an important role in solving this problem. In the contexts of both the state-space and singular systems, a great number of results on H_∞ control have been reported and different approaches have been proposed in the literature [41, 80, 88, 125, 191].

In this chapter, we deal with the H_∞ control problem for singular systems in both the continuous- and discrete-time contexts. The purpose is to design a controller such that the closed-loop system is regular, impulse-free (for continuous singular systems) or causality (for discrete singular systems), stable and the transfer function of the closed-loop system satisfies a prescribed H_∞ performance level. To solve this problem, we first present versions of the bounded real lemma for singular systems. Based on these results, necessary and sufficient conditions for the solvability of the H_∞ control problem are obtained. Both state feedback and dynamic output feedback controllers are designed for continuous singular systems. When parameter uncertainties appear in the system model, the concept of generalized quadratic stability with an H_∞-norm bound is proposed, which implies the existence of a state feedback controller such that the closed-loop system is admissible and the transfer function satisfies a prescribed H_∞ performance level for all allowable uncertainties. Necessary and sufficient conditions for generalized quadratic stabilizability with an H_∞-norm bound are derived. When these conditions are feasible, desired state feedback controllers are also constructed.

5.2 Continuous Systems

In this section, we consider the H_∞ control problem for continuous singular systems. Versions of the bounded real lemma are proposed via LMIs. With these results, state feedback and dynamic output feedback controllers will be considered to solve the H_∞ control problem.

5.2.1 State Feedback Control

Here, attention will be focused on the design of state feedback controllers to solve the H_∞ control problem. The class of continuous singular systems to be considered are described by

$$E\dot{x}(t) = Ax(t) + B_1 u(t) + B\omega(t), \tag{5.1}$$
$$z(t) = Cx(t) + B_2 u(t), \tag{5.2}$$
$$y(t) = Hx(t) + B_3 u(t), \tag{5.3}$$

where $x(t) \in \mathbb{R}^n$ is the state; $u(t) \in \mathbb{R}^m$ is the control input; $y(t) \in \mathbb{R}^l$ is the measurement; $\omega(t) \in \mathbb{R}^p$ is the disturbance input which belongs to $\mathcal{L}_2[0, \infty)$, and $z(t) \in \mathbb{R}^s$ is the controlled output. The matrix $E \in \mathbb{R}^{n \times n}$ may be singular; we shall assume that $\text{rank}(E) = r \leq n$. A, B, B_1, B_2, B_3, C and H are known real constant matrices with appropriate dimensions.

It is known that the bounded real lemma has played an important role in the study of the H_∞ control problem for state-space systems. Considering this, we first present a version of bounded real lemma for the continuous singular system in (5.1) and (5.2) before investigating the H_∞ control problem. To this end, we consider the unforced singular system of (5.1) and (5.2); that is,

$$E\dot{x}(t) = Ax(t) + B\omega(t), \tag{5.4}$$
$$z(t) = Cx(t). \tag{5.5}$$

For this continuous singular system, we have the following bounded real lemma.

Theorem 5.1. *Given a scalar $\gamma > 0$. The continuous singular system in (5.4) and (5.5) is admissible and its transfer function*

$$G(s) = C(sE - A)^{-1} B, \tag{5.6}$$

satisfies

$$\|G\|_\infty < \gamma, \tag{5.7}$$

if and only if there exists a matrix P such that the following LMIs hold:

$$E^T P = P^T E \geq 0, \tag{5.8}$$

$$\begin{bmatrix} A^T P + P^T A + C^T C & P^T B \\ B^T P & -\gamma^2 I \end{bmatrix} < 0. \tag{5.9}$$

To prove this theorem, we need the bounded real lemma for continuous state-space systems, which is stated in the following lemma.

Lemma 5.1. [55] *Given a scalar $\gamma > 0$. The continuous state-space system described by*

$$\dot{x}(t) = Ax(t) + B\omega(t),$$
$$z(t) = Cx(t) + D\omega(t),$$

is stable and its transfer function

$$\hat{G}(s) = C\left(sE - A\right)^{-1} B + D,$$

satisfies

$$\|\hat{G}\|_\infty < \gamma,$$

if and only if there exists a matrix $P > 0$ such that the following LMI holds:

$$\begin{bmatrix} A^T P + P^T A + C^T C & PB + C^T D \\ B^T P + D^T C & D^T D - \gamma^2 I \end{bmatrix} < 0.$$

Proof of Theorem 5.1. (*Sufficiency*) Suppose that there exists a matrix P such that the LMIs in (5.8) and (5.9) hold. Then, it is easy to see that

$$A^T P + P^T A < 0.$$

Noting this, (5.8), and then using Theorem 2.1, we have that the continuous singular system in (5.4) is admissible. In the following, we will show that the transfer function $G(s)$ in (5.6) satisfies (5.7). To this end, we apply Schur complement to (5.9) and obtain

$$A^T P + P^T A + C^T C + \gamma^{-2} P^T B B^T P < 0, \tag{5.10}$$

which implies that there exists a matrix $\Gamma > 0$ such that

$$A^T P + P^T A + C^T C + \gamma^{-2} P^T B B^T P + \Gamma < 0.$$

With this and (5.8), it can be seen that

$$(-j\theta E - A)^T P + P^T \left(j\theta E - A\right) - C^T C - \Omega > 0, \tag{5.11}$$

for all $\theta \in \mathbb{R}$, where

$$\Omega = \gamma^{-2} P^T BB^T P + \Gamma.$$

Since the continuous singular system in (5.4) and (5.5) is admissible, $(j\theta E - A)^{-1}$ is well defined. Thus, pre- and post-multiplying (5.11) by $B^T (-j\theta E - A)^{-T}$ and $(j\theta E - A)^{-1} B$, respectively, we have that for all $\theta \in \mathbb{R}$,

$$B^T P (j\theta E - A)^{-1} B + B^T (-j\theta E - A)^{-T} P^T$$
$$-B^T (-j\theta E - A)^{-T} C^T C (j\theta E - A)^{-1} B$$
$$-B^T (-j\theta E - A)^{-T} \Omega (j\theta E - A)^{-1} B \geq 0.$$

Using this, we obtain that for all $\theta \in \mathbb{R}$,

$$\gamma^2 I - B^T (-j\theta E - A)^{-T} C^T C (j\theta E - A)^{-1} B$$
$$\geq \gamma^2 I - B^T P (j\theta E - A)^{-1} B - B^T (-j\theta E - A)^{-T} P^T B$$
$$+B^T (-j\theta E - A)^{-T} \Omega (j\theta E - A)^{-1} B$$
$$= \gamma^2 I + \left[B^T (-j\theta E - A)^{-T} - B^T P \Omega^{-1} \right] \Omega \left[(j\theta E - A)^{-1} B - \Omega^{-1} P^T B \right]$$
$$-B^T P \Omega^{-1} P^T B$$
$$\geq \gamma^2 I - B^T P \Omega^{-1} P^T B. \tag{5.12}$$

Note that

$$\Omega - \gamma^{-2} P^T BB^T P = \Gamma > 0,$$

which, by Schur complement, gives

$$\begin{bmatrix} \Omega & P^T B \\ B^T P & \gamma^2 I \end{bmatrix} > 0.$$

Then, applying Schur complement to this inequality, we have

$$\gamma^2 I - B^T P \Omega^{-1} P^T B > 0.$$

This together with (5.12) implies

$$\gamma^2 I - B^T (-j\theta E - A)^{-T} C^T C (j\theta E - A)^{-1} B > 0,$$

for all $\theta \in \mathbb{R}$. That is, (5.7) is satisfied.

(*Necessity*) Suppose that the continuous singular system in (5.4) and (5.5) is admissible and its transfer function $G(s)$ in (5.6) satisfies (5.7). Then, there exist nonsingular matrices M and N such that

$$E = M \begin{bmatrix} I & 0 \\ 0 & 0 \end{bmatrix} N, \quad A = M \begin{bmatrix} A_1 & 0 \\ 0 & I \end{bmatrix} N, \tag{5.13}$$

where $A_1 \in \mathbb{R}^{r \times r}$ is stable. Write

$$CN^{-1} = \begin{bmatrix} \tilde{C}_1 & \tilde{C}_2 \end{bmatrix}, \quad M^{-1}B = \begin{bmatrix} \tilde{B}_1 \\ \tilde{B}_2 \end{bmatrix}.$$

where the partition is compatible with that of A in (5.13). Then, it is easy to see that

$$G(s) = \tilde{C}_1 \left(sI - A_1 \right)^{-1} \tilde{B}_1 - \tilde{C}_2 \tilde{B}_2.$$

Then, by Lemma 5.1, we have that the stability of A_1 and the inequality in (5.7) imply that there exists a matrix $P_1 > 0$ such that

$$\begin{bmatrix} A_1^T P_1 + P_1 A_1 + \tilde{C}_1^T \tilde{C}_1 & P_1 \tilde{B}_1 - \tilde{C}_1^T \tilde{C}_2 \tilde{B}_2 \\ \tilde{B}_1^T P_1 - \tilde{B}_2^T \tilde{C}_2^T \tilde{C}_1 & \gamma^2 I - \tilde{B}_2^T \tilde{C}_2^T \tilde{C}_2 \tilde{B}_2 \end{bmatrix} < 0.$$

With this, it is easy to see that there exists a scalar $\varepsilon > 0$ satisfying

$$\begin{bmatrix} A_1^T P_1 + P_1 A_1 + \tilde{C}_1^T \tilde{C}_1 & P_1 \tilde{B}_1 - \tilde{C}_1^T \tilde{C}_2 \tilde{B}_2 \\ \tilde{B}_1^T P_1 - \tilde{B}_2^T \tilde{C}_2^T \tilde{C}_1 & \tilde{B}_2^T \left(\tilde{C}_2^T \tilde{C}_2 + \varepsilon I \right) \tilde{B}_2 - \gamma^2 I \end{bmatrix} < 0. \tag{5.14}$$

Let

$$\Psi = \tilde{C}_2^T \tilde{C}_2 + \varepsilon I.$$

Then, by (5.14), we have

$$\begin{bmatrix} A_1^T P_1 + P_1 A_1 + \tilde{C}_1^T \tilde{C}_1 & P_1 \tilde{B}_1 - \tilde{C}_1^T \tilde{C}_2 \tilde{B}_2 \\ \tilde{B}_1^T P_1 - \tilde{B}_2^T \tilde{C}_2^T \tilde{C}_1 & \tilde{B}_2^T \Psi \left(\tilde{C}_2^T \tilde{C}_2 + 2\varepsilon I \right)^{-1} \Psi \tilde{B}_2 - \gamma^2 I \end{bmatrix} < 0,$$

which, by Schur complement, provides

$$\begin{bmatrix} A_1^T P_1 + P_1 A_1 + \tilde{C}_1^T \tilde{C}_1 & P_1 \tilde{B}_1 - \tilde{C}_1^T \tilde{C}_2 \tilde{B}_2 & 0 \\ \tilde{B}_1^T P_1 - \tilde{B}_2^T \tilde{C}_2^T \tilde{C}_1 & -\gamma^2 I & \tilde{B}_2^T \Psi \\ 0 & \Psi \tilde{B}_2 & \tilde{C}_2^T \tilde{C}_2 - 2\Psi \end{bmatrix} < 0.$$

Therefore,

$$\begin{bmatrix} A_1^T P_1 + P_1 A_1 + \tilde{C}_1^T \tilde{C}_1 & 0 & P_1 \tilde{B}_1 - \tilde{C}_1^T \tilde{C}_2 \tilde{B}_2 \\ 0 & \tilde{C}_2^T \tilde{C}_2 - 2\Psi & -\Psi \tilde{B}_2 \\ \tilde{B}_1^T P_1 - \tilde{B}_2^T \tilde{C}_2^T \tilde{C}_1 & -\tilde{B}_2^T \Psi & -\gamma^2 I \end{bmatrix} < 0. \tag{5.15}$$

Now, set

$$P = M^{-T} \begin{bmatrix} P_1 & 0 \\ -\tilde{C}_2^T \tilde{C}_1 & -\Psi \end{bmatrix} N. \tag{5.16}$$

Then, by (5.15), it can be verified that the matrix P given in (5.16) satisfies

$$\begin{bmatrix} N^{-T} \left(A^T P + P^T A + C^T C \right) N^{-1} & N^{-T} P^T B \\ B^T P N^{-1} & -\gamma^2 I \end{bmatrix} < 0.$$

Pre- and post-multiplying this inequality by $\operatorname{diag}(N^T, I)$ and its transpose, respectively, provide (5.9). Therefore, the matrix P in (5.16) satisfies (5.8) and (5.9). This completes the proof.   □

Theorem 5.1 can be viewed as an extension of the bounded real lemma for continuous state-space systems to continuous singular systems. Note that

$$\|G\|_\infty = \|G^T\|_\infty.$$

Thus, by Theorem 5.1, it is easy to obtain the following corollary.

Corollary 5.1. *Given a scalar* $\gamma > 0$. *The continuous singular system in (5.4) and (5.5) is admissible and its transfer function* $G(s)$ *in (5.6) satisfies (5.7) if and only if there exists a matrix* P *such that the following LMIs hold:*

$$EP = P^T E^T \geq 0, \tag{5.17}$$

$$\begin{bmatrix} P^T A^T + AP + BB^T & P^T C^T \\ CP & -\gamma^2 I \end{bmatrix} < 0. \tag{5.18}$$

It is noted that both Theorem 5.1 and Corollary 5.1 involve non-strict LMIs. Similar to the derivation of Theorem 2.2, we can obtain the following version of the bounded real lemma for continuous singular systems in terms of a strict LMI.

Theorem 5.2. *Given a scalar* $\gamma > 0$. *The continuous singular system in (5.4) and (5.5) is admissible and its transfer function* $G(s)$ *in (5.6) satisfies (5.7) if and only if there exist matrices* $P > 0$ *and* Q *such that the following LMI holds:*

$$\begin{bmatrix} (PE + SQ)^T A + A^T (PE + SQ) + C^T C & (PE + SQ)^T B \\ B^T (PE + SQ) & -\gamma^2 I \end{bmatrix} < 0, \tag{5.19}$$

where $S \in \mathbb{R}^{n \times (n-r)}$ *is any matrix with full column rank and satisfies* $E^T S = 0$.

Similarly, by Corollaries 2.2 and 5.1, it is easy to obtain the following corollary.

Corollary 5.2. *Given a scalar* $\gamma > 0$. *The continuous singular system in (5.4) and (5.5) is admissible and its transfer function* $G(s)$ *in (5.6) satisfies (5.7) if and only if there exist matrices* $P > 0$ *and* Q *such that the following LMI holds:*

$$\left[\begin{array}{cc} \left(PE^T + SQ\right)^T A^T + A\left(PE^T + SQ\right) + BB^T & \left(PE^T + SQ\right)^T C^T \\ C\left(PE^T + SQ\right) & -\gamma^2 I \end{array} \right] < 0.$$

$$(5.20)$$

where $S \in \mathbb{R}^{n \times (n-r)}$ is any matrix with full column rank and satisfies $ES = 0$.

Now, consider the following state feedback controller

$$u(t) = Kx(t), \quad K \in \mathbb{R}^{m \times n}. \tag{5.21}$$

Applying this controller to (5.1) and (5.2) results in the following closed-loop system:

$$E\dot{x}(t) = (A + B_1 K)x(t) + B\omega(t), \tag{5.22}$$
$$z(t) = (C + B_2 K)x(t). \tag{5.23}$$

Then, we have the following H_∞ control result. The proof can be carried out by resorting to Corollary 5.2 and following a similar line as in the proof of Theorem 3.1, and is thus omitted.

Theorem 5.3. *Given a scalar $\gamma > 0$ and the continuous singular system in (5.1) and (5.2). There exists a state feedback controller in (5.21) such that the closed-loop system in (5.22) and (5.23) is admissible and its transfer function*

$$G_{cK}(s) = (C + B_2 K)\left[sI - (A + B_1 K)\right]^{-1} B,$$

satisfies

$$\|G_{cK}\|_\infty < \gamma,$$

if and only if there exist matrices $P > 0$, Q and Y such that

$$\left[\begin{array}{cc} \Pi(P,Q) + BB^T & \Omega(P,Q)^T C^T + Y^T B_2^T \\ C\Omega(P,Q) + B_2 Y & -\gamma^2 I \end{array} \right] < 0. \tag{5.24}$$

where

$$\Pi(P,Q) = \Omega(P,Q)^T A^T + A\Omega(P,Q) + B_1 Y + Y^T B_1^T, \tag{5.25}$$
$$\Omega(P,Q) = PE^T + SQ, \tag{5.26}$$

and $S \in \mathbb{R}^{n \times (n-r)}$ is any matrix with full column rank and satisfies $ES = 0$. In this case, we can assume that the matrix $\Omega(P,Q)$ is nonsingular (if this is not the case, then we can choose some $\theta \in (0,1)$ such that $\hat{\Omega}(P,Q) = \Omega(P,Q) + \theta \tilde{P}$ is nonsingular and satisfies (5.24), in which $\tilde{P}$ is any nonsingular matrix satisfying $E\tilde{P} = \tilde{P}^T E^T \geq 0$), then a desired state feedback controller can be chosen as

$$u(t) = Y\Omega(P,Q)^{-1} x(t).$$

5.2.2 Output Feedback Control

Now, we consider the H_∞ problem for continuous singular systems via dynamic output feedback controllers. The system to be considered is given in (5.1)–(5.3). For this singular system, we consider the following dynamic output feedback controller:

$$E_K\dot{\xi}(t) = A_K\xi(t) + B_Ky(t), \tag{5.27}$$
$$u(t) = C_K\xi(t), \tag{5.28}$$

where $\xi(t) \in \mathbb{R}^n$ is the controller state, E_K, A_K, B_K and C_K are constant matrices to be determined. Applying this controller to (5.1)–(5.3), we obtain the following closed-loop system:

$$E_c\dot{\eta}(t) = A_c\eta(t) + B_c\omega(t), \tag{5.29}$$
$$z(t) = C_c\eta(t), \tag{5.30}$$

where

$$\eta(t) = \begin{bmatrix} x(t) \\ \xi(t) \end{bmatrix}, \tag{5.31}$$

and

$$E_c = \begin{bmatrix} E & 0 \\ 0 & E_K \end{bmatrix}, \tag{5.32}$$

$$A_c = \begin{bmatrix} A & B_1C_K \\ B_KH & A_K + B_KB_3C_K \end{bmatrix}, \tag{5.33}$$

$$B_c = \begin{bmatrix} B \\ 0 \end{bmatrix}, \tag{5.34}$$

$$C_c = \begin{bmatrix} C & B_2C_K \end{bmatrix}. \tag{5.35}$$

Then, we have the following H_∞ control result.

Theorem 5.4. *Given a scalar $\gamma > 0$ and the continuous singular system in (5.1)–(5.3). There exists a dynamic output feedback controller in (5.27) and (5.28) such that the closed-loop system in (5.29) and (5.30) is admissible and its transfer function*

$$G_c(s) = C_c\left(sI - A_c\right)^{-1}B_c,$$

satisfies

$$\|G_c\|_\infty < \gamma, \tag{5.36}$$

if and only if there exist matrices X, Y, Φ and Ψ such that

$$\begin{bmatrix} E^T & 0 \\ 0 & E \end{bmatrix} \begin{bmatrix} X & I \\ I & Y \end{bmatrix} = \begin{bmatrix} X^T & I \\ I & Y^T \end{bmatrix} \begin{bmatrix} E & 0 \\ 0 & E^T \end{bmatrix} \geq 0, \quad (5.37)$$

$$\begin{bmatrix} A^T X + X^T A + \Phi H + H^T \Phi^T & X^T B & C^T \\ BX & -\gamma^2 I & 0 \\ C & 0 & -I \end{bmatrix} < 0, \quad (5.38)$$

$$\begin{bmatrix} AY + Y^T A^T - B_1 \Psi - \Psi^T B_1^T & B & Y^T C^T - \Psi^T B_2^T \\ B^T & -\gamma^2 I & 0 \\ CY - B_2 \Psi & 0 & -I \end{bmatrix} < 0. \quad (5.39)$$

When the LMIs in (5.37)–(5.39) are feasible, we can always find matrices X, Y, Φ and Ψ that satisfy (5.37)–(5.39) and both Y and $Y^{-1} - X$ are nonsingular, and a desired dynamic output feedback controller in (5.27) and (5.28) can be chosen with the following parameters:

$$A_K = \left(Y^{-1} - X\right)^{-T} \left[\left(X^T B_1 + \Phi B_3 + C^T B_2\right) \Psi Y^{-1} \right.$$
$$\left. - X^T A - \Phi H - \gamma^{-2} X^T B B^T Y^{-1} - C^T C - A^T Y^{-1}\right], \quad (5.40)$$

$$E_K = E, \quad (5.41)$$

$$B_K = \left(Y^{-1} - X\right)^{-T} \Phi, \quad (5.42)$$

$$C_K = -\Psi Y^{-1}. \quad (5.43)$$

Proof. (Sufficiency) Similar to the proof of Theorem 3.2, it can be shown that when (5.37)–(5.39) are feasible, there always exist matrices X, Y, Φ and Ψ that satisfy (5.37)–(5.39) and both Y and $Y^{-1} - X$ are nonsingular. Now, with the dynamic output feedback controller with parameters given in (5.40)–(5.43), we have the following closed-loop system:

$$\tilde{E}_c \dot{\eta}(t) = \tilde{A}_c \eta(t) + \tilde{B}_c \omega(t), \quad (5.44)$$
$$z(t) = \tilde{C}_c \eta(t), \quad (5.45)$$

where $\eta(t)$ is given in (5.31), and

$$\tilde{E}_c = \begin{bmatrix} E & 0 \\ 0 & E \end{bmatrix},$$
$$\tilde{A}_c = \begin{bmatrix} A & -B_1 \Psi Y^{-1} \\ \left(Y^{-1} - X\right)^{-T} \Phi H & \tilde{\Lambda} \end{bmatrix},$$
$$\tilde{B}_c = \begin{bmatrix} B \\ 0 \end{bmatrix},$$
$$\tilde{C}_c = \begin{bmatrix} C & -B_2 \Psi Y^{-1} \end{bmatrix},$$

where

$$\tilde{\Lambda} = \left(Y^{-1} - X\right)^{-T} \left[\left(X^T B_1 + C^T B_2\right) \Psi Y^{-1} \right.$$
$$\left. - X^T A - \Phi H - \gamma^{-2} X^T B B^T Y^{-1} - C^T C - A^T Y^{-1}\right].$$

Let

$$\tilde{P}_c = \begin{bmatrix} X & Y^{-1} - X \\ Y^{-1} - X & X - Y^{-1} \end{bmatrix}.$$

Then, similar to the proof of Theorem 3.2, we can verify that

$$\tilde{E}_c^T \tilde{P}_c = \tilde{P}_c^T \tilde{E}_c \geq 0. \tag{5.46}$$

Now, let

$$\begin{aligned}
\Xi_1 &= A^T X + X^T A + \Phi H + H^T \Phi^T + \gamma^{-2} X^T B B^T X + C^T C, \\
\Xi_2 &= A Y + Y^T A^T - B_1 \Psi - \Psi^T B_1^T + \gamma^{-2} B B^T \\
&\quad + \left(Y^T C^T - \Psi^T B_2^T \right) \left(C Y - B_2 \Psi \right).
\end{aligned}$$

Then, by applying Schur complement to (5.38) and (5.39), we have

$$\Xi_1 < 0, \tag{5.47}$$
$$\Xi_2 < 0. \tag{5.48}$$

By algebraic manipulations, it can be verified that

$$\Upsilon^T \left[\tilde{A}_c^T \tilde{P}_c + \tilde{P}_c^T \tilde{A}_c + \gamma^{-2} \tilde{P}_c^T \tilde{B}_c \tilde{B}_c^T \tilde{P}_c + \tilde{C}_c^T \tilde{C}_c \right] \Upsilon = \begin{bmatrix} \Xi_1 & 0 \\ 0 & \Xi_2 \end{bmatrix} \tag{5.49}$$

where

$$\Upsilon = \begin{bmatrix} I & Y \\ 0 & Y \end{bmatrix}.$$

It follows from (5.47)–(5.49) that

$$\Upsilon^T \left[\tilde{A}_c^T \tilde{P}_c + \tilde{P}_c^T \tilde{A}_c + \gamma^{-2} \tilde{P}_c^T \tilde{B}_c \tilde{B}_c^T \tilde{P}_c + \tilde{C}_c^T \tilde{C}_c \right] \Upsilon < 0.$$

Pre- and post-multiplying this inequality by Υ^{-T} and its transpose result in

$$\tilde{A}_c^T \tilde{P}_c + \tilde{P}_c^T \tilde{A}_c + \gamma^{-2} \tilde{P}_c^T \tilde{B}_c \tilde{B}_c^T \tilde{P}_c + \tilde{C}_c^T \tilde{C}_c < 0,$$

which, by Schur complement, implies

$$\begin{bmatrix} \tilde{A}_c^T \tilde{P}_c + \tilde{P}_c^T \tilde{A}_c & \tilde{P}_c^T \tilde{B}_c & \tilde{C}_c^T \\ \tilde{B}_c^T \tilde{P}_c & -\gamma^2 I & 0 \\ \tilde{C}_c & 0 & -I \end{bmatrix} < 0.$$

Therefore, with this and Theorem 5.1, we have that the closed-loop system in (5.44) and (5.45) is admissible and its transfer function satisfies (5.36).

(*Necessity*) Suppose that there exists a dynamic output feedback controller in (5.27) and (5.28) such that the closed-loop system in (5.29) and (5.30) is admissible and its transfer function satisfies (5.36). Then, it follows from Theorem 5.1 that there exists a matrix P_c such that

$$E_c^T P_c = P_c^T E_c \geq 0, \tag{5.50}$$

$$\begin{bmatrix} A_c^T P_c + P_c^T A_c & P_c^T B_c & C_c^T \\ B_c^T P_c & -\gamma^2 I & 0 \\ C_c & 0 & -I \end{bmatrix} < 0. \tag{5.51}$$

Write

$$P_c = \begin{bmatrix} P_{c1} & P_{c2} \\ P_{c3} & P_{c4} \end{bmatrix},$$

where the partition is compatible with that of E_c in (5.32). Then, along a similar line as in the proof of Theorem 3.2, without loss of generality, we can assume that both P_{c4} and $P_{c1} - P_{c2}P_{c4}^{-1}P_{c3}$ are nonsingular. Substituting the parameters in (5.32)–(5.35) to (5.51), we have

$$\begin{bmatrix} P_{c1}^T A + A^T P_{c1} + P_{c3}^T B_k H + H^T B_k^T P_{c3} & \Theta_1 & P_{c1}^T B & C^T \\ \Theta_1^T & \Theta_2 & P_{c2}^T B & C_K^T B_2^T \\ B^T P_{c1} & B^T P_{c2} & -\gamma^2 I & 0 \\ C & B_2 C_K & 0 & -I \end{bmatrix} < 0, \tag{5.52}$$

where

$$\Theta_1 = P_{c1}^T B_1 C_K + P_{c3}^T (A_K + B_K B_3 C_K) + A^T P_{c2} + H^T B_K^T P_{c4},$$
$$\Theta_2 = P_{c2}^T B_1 C_K + C_K^T B_1^T P_{c2} + P_{c4}^T (A_K + B_K B_3 C_K)$$
$$+ (A_K + B_K B_3 C_K)^T P_{c4}.$$

Pre- and post-multiplying (5.52) by

$$\begin{bmatrix} I & 0 & 0 & 0 \\ 0 & 0 & I & 0 \\ 0 & 0 & 0 & I \end{bmatrix}$$

and its transpose give

$$\begin{bmatrix} P_{c1}^T A + A^T P_{c1} + P_{c3}^T B_k H + H^T B_k^T P_{c3} & P_{c1}^T B & C^T \\ B^T P_{c1} & -\gamma^2 I & 0 \\ C & 0 & -I \end{bmatrix} < 0. \tag{5.53}$$

Now, let

$$X = P_{c1}, \quad \Phi = P_{c3}^T B_K.$$

Then, it is easy to see that (5.53) provides (5.38). Letting

$$U = \begin{bmatrix} I & 0 \\ -P_{c4}^{-1} P_{c3} & I \end{bmatrix},$$

and pre- and post-multiplying (5.51) by

$$\begin{bmatrix} U^T & 0 & 0 \\ 0 & I & 0 \\ 0 & 0 & I \end{bmatrix}$$

and its transpose, respectively, we obtain

$$\begin{bmatrix} \Lambda & * & W^T B & C^T - P_{c3}^T P_{c4}^{-T} C_K^T B_2^T \\ * & * & * & * \\ B^T W & * & -\gamma^2 I & 0 \\ C - B_2 C_K P_{c4}^{-1} P_{c3} & * & 0 & -I \end{bmatrix} < 0, \qquad (5.54)$$

where

$$W = P_{c1} - P_{c2} P_{c4}^{-1} P_{c3},$$
$$\Lambda = W^T A + A^T W - W^T B_1 C_K P_{c4}^{-1} P_{c3} - P_{c3}^T P_{c4}^{-T} C_K^T B_1^T W,$$

and $*$ represents a matrix which will not be used in the following discussion. Now, pre- and post-multiplying (5.54) by

$$\begin{bmatrix} W^{-T} & 0 & 0 & 0 \\ 0 & 0 & I & 0 \\ 0 & 0 & 0 & I \end{bmatrix}$$

and its transpose give

$$\begin{bmatrix} W^{-T} \Lambda W^{-1} & B & W^{-T}\left(C^T - P_{c3}^T P_{c4}^{-T} C_K^T B_2^T\right) \\ B^T & -\gamma^2 I & 0 \\ \left(C - B_2 C_K P_{c4}^{-1} P_{c3}\right) W^{-1} & 0 & -I \end{bmatrix} < 0.$$
$$(5.55)$$

Set

$$Y = W^{-1}, \quad \Psi = C_K P_{c4}^{-1} P_{c3} Y.$$

Then, it is easy to see that (5.55) provides (5.39). Finally, following a similar line as in the proof of Theorem 3.2, we can show that (5.50) implies (5.37). This completes the proof. □

5.3 Discrete Systems

In this section, we investigate the H_∞ control problem for discrete singular systems. Similar to the continuous case, we first propose versions of the bounded real lemma for discrete singular systems, and then the H_∞ control problem is solved via state feedback controllers.

Consider a linear discrete singular system described by

$$Ex(k+1) = Ax(k) + B_1 u(k) + B\omega(k), \qquad (5.56)$$
$$z(k) = Cx(k) + D\omega(k), \qquad (5.57)$$

where $x(k) \in \mathbb{R}^n$ is the state; $u(k) \in \mathbb{R}^m$ is the control input; $w(k) \in \mathbb{R}^p$ is the disturbance input which belongs to $l_2[0, \infty)$, and $z(k) \in \mathbb{R}^s$ is the controlled output. The matrix $E \in \mathbb{R}^{n \times n}$ may be singular; we shall assume that $\text{rank}(E) = r \leq n$. A, B, B_1, C and D are known real constant matrices with appropriate dimensions.

Firstly, we present versions of the bounded real lemma for the discrete singular system in (5.56) and (5.57). To this end, we consider the unforced singular system of (5.56) and (5.57); that is,

$$Ex(k+1) = Ax(k) + Bw(k), \tag{5.58}$$
$$z(k) = Cx(k) + Dw(k). \tag{5.59}$$

Then, for the discrete singular system in (5.58) and (5.59), we have the following bounded real lemma. The result has been given in [80] but a different proof is offered here.

Theorem 5.5. *Given a scalar $\gamma > 0$. The discrete singular system in (5.58) and (5.59) is admissible and its transfer function*

$$G(z) = C\left(zE - A\right)^{-1} B + D, \tag{5.60}$$

satisfies

$$\|G\|_\infty < \gamma, \tag{5.61}$$

if and only if there exists a matrix $P = P^T$ such that the following LMIs hold:

$$E^T P E \geq 0, \tag{5.62}$$

$$\begin{bmatrix} A^T P A - E^T P E + C^T C & A^T P B + C^T D \\ B^T P A + D^T C & B^T P B + D^T D - \gamma^2 I \end{bmatrix} < 0. \tag{5.63}$$

To prove this theorem, we need the following lemmas.

Lemma 5.2. [55] *Given a scalar $\gamma > 0$. The discrete state-space system described by*

$$x(k+1) = Ax(k) + Bw(k),$$
$$z(k) = Cx(k) + Dw(k),$$

is stable and its transfer function

$$\hat{G}(z) = C\left(zI - A\right)^{-1} B + D,$$

satisfies

$$\|\hat{G}\|_\infty < \gamma,$$

if and only if there exists a matrix $P > 0$ such that the following LMI holds:

$$\begin{bmatrix} A^T P A + C^T C - P & A^T P B + C^T D \\ B^T P A + D^T C & B^T P B + D^T D - \gamma^2 I \end{bmatrix} < 0.$$

Lemma 5.3. *Given a scalar $\gamma > 0$. The discrete singular system*

$$Ex(k+1) = Ax(k) + B\omega(k), \tag{5.64}$$
$$z(k) = Cx(k), \tag{5.65}$$

is admissible and its transfer function

$$G_1(z) = C\left(zE - A\right)^{-1} B, \tag{5.66}$$

satisfies

$$\|G_1\|_\infty < \gamma, \tag{5.67}$$

if and only if there exists a matrix $P = P^T$ such that the following LMIs hold:

$$E^T P E \geq 0, \tag{5.68}$$

$$\begin{bmatrix} A^T P A - E^T P E + C^T C & A^T P B \\ B^T P A & B^T P B - \gamma^2 I \end{bmatrix} < 0. \tag{5.69}$$

Proof. (*Sufficiency*) Suppose that there exists a matrix $P = P^T$ such that the LMIs in (5.68) and (5.69) hold. Then, it is easy to see that

$$A^T P A - E^T P E < 0.$$

Noting this inequality, (5.68), and using Theorem 2.3, we have that the discrete singular system in (5.64) is admissible. Next, we will show that the transfer function (5.66) satisfies (5.67). To this end, we apply Schur complement to (5.69) and obtain

$$A^T P A - E^T P E + C^T C$$
$$+ A^T P B \left(\gamma^2 I - B^T P B\right)^{-1} B^T P A < 0,$$

which implies that there exists a matrix $\Xi > 0$ such that

$$A^T P A - E^T P E + C^T C + \Pi < 0, \tag{5.70}$$

where

$$\Pi = A^T P B \left(\gamma^2 I - B^T P B\right)^{-1} B^T P A + \Xi.$$

Pre- and post-multiplying (5.70) by $B^T \left(e^{-j\theta} E^T - A^T\right)^{-1}$ and $\left(e^{j\theta} E - A\right)^{-1} B$, respectively, we obtain

$$B^T \left(e^{-j\theta} E^T - A^T\right)^{-1} \left(A^T PA - E^T PE + C^T C + \Pi\right) \left(e^{j\theta} E - A\right)^{-1} B \le 0. \tag{5.71}$$

Observe that

$$B^T \left(e^{-j\theta} E^T - A^T\right)^{-1} \left(A^T PA - E^T PE\right) \left(e^{j\theta} E - A\right)^{-1} B$$
$$= -B^T PB - B^T PA \left(e^{j\theta} E - A\right)^{-1} B - B^T \left(e^{-j\theta} E^T - A^T\right)^{-1} A^T PB.$$

From this and (5.71), it follows that

$$\gamma^2 I - B^T \left(e^{-j\theta} E^T - A^T\right)^{-1} C^T C \left(e^{j\theta} E - A\right)^{-1} B$$
$$\ge \gamma^2 I - B^T PB + B^T \left(e^{-j\theta} E^T - A^T\right)^{-1} \Pi \left(e^{j\theta} E - A\right)^{-1} B$$
$$- B^T PA \left(e^{j\theta} E - A\right)^{-1} B - B^T \left(e^{-j\theta} E^T - A^T\right)^{-1} A^T PB. \tag{5.72}$$

Note that

$$B^T \left(e^{-j\theta} E^T - A^T\right)^{-1} \Pi \left(e^{j\theta} E - A\right)^{-1} B$$
$$- B^T PA \left(e^{j\theta} E - A\right)^{-1} B - B^T \left(e^{-j\theta} E^T - A^T\right)^{-1} A^T PB$$
$$= \left[B^T \left(e^{-j\theta} E^T - A^T\right)^{-1} - B^T PA\Pi^{-1}\right]$$
$$\times \Pi \left[\left(e^{j\theta} E - A\right)^{-1} B - \Pi^{-1} A^T PB\right] - B^T PA\Pi^{-1} A^T PB$$
$$\ge -B^T PA\Pi^{-1} A^T PB.$$

This together with (5.72) implies

$$\gamma^2 I - B^T \left(e^{-j\theta} E^T - A^T\right)^{-1} C^T C \left(e^{j\theta} E - A\right)^{-1} B$$
$$\ge \gamma^2 I - B^T PB - B^T PA\Pi^{-1} A^T PB. \tag{5.73}$$

Now, we note

$$\Pi - A^T PB \left(\gamma^2 I - B^T PB\right)^{-1} B^T PA = \Xi > 0.$$

By Schur complement, it is easy to have

$$\begin{bmatrix} \Pi & A^T PB \\ B^T PA & \gamma^2 I - B^T PB \end{bmatrix} > 0,$$

which, by Schur complement again, provides

$$\gamma^2 I - B^T PB - B^T PA\Pi^{-1} A^T PB > 0.$$

With this and (5.73), we have that the transfer function (5.66) satisfies (5.67).

(*Necessity*) Suppose that the discrete singular system in (5.64) and (5.65) is admissible and its transfer function (5.66) satisfies (5.67). Then, it follows from Lemmas 2.9 and 2.10 that there exist two nonsingular matrices L_1 and L_2 such that

$$L_1 E L_2 = \begin{bmatrix} I & 0 \\ 0 & 0 \end{bmatrix}, \quad L_1 A L_2 = \begin{bmatrix} A_1 & 0 \\ 0 & I \end{bmatrix}. \tag{5.74}$$

where $A_1 \in \mathbb{R}^{r \times r}$ and

$$\rho(A_1) < 1.$$

Write

$$L_1 B = \begin{bmatrix} \tilde{B}_1 \\ \tilde{B}_2 \end{bmatrix}, \quad C L_2 = [\tilde{C}_1 \ \tilde{C}_2], \tag{5.75}$$

where the partition is compatible with that of E in (5.74). Then, it is easy to have

$$G_1(z) = \tilde{C}_1 (zE - A_1)^{-1} \tilde{B}_1 - \tilde{C}_2 \tilde{B}_2.$$

Noting this, (5.67), and then applying Lemma 5.2, we have that there exists a matrix $Q > 0$ such that

$$\begin{bmatrix} A_1^T Q A_1 + \tilde{C}_1^T \tilde{C}_1 - Q & A_1^T Q \tilde{B}_1 - \tilde{C}_1^T \tilde{C}_2 \tilde{B}_2 \\ \tilde{B}_1^T Q A - \tilde{B}_2^T \tilde{C}_2^T \tilde{C}_1 & \tilde{B}_1^T Q \tilde{B}_1 + \tilde{B}_2^T \tilde{C}_2^T \tilde{C}_2 \tilde{B}_2 - \gamma^2 I \end{bmatrix} < 0,$$

which, by Schur complement, implies

$$\Delta + \left(\tilde{C}_1^T \tilde{C}_2 \tilde{B}_2 - \bar{A} \right) M^{-1} \left(\tilde{C}_1^T \tilde{C}_2 \tilde{B}_2 - \bar{A} \right) < 0, \tag{5.76}$$

where

$$\begin{aligned} \Delta &= A_1^T Q A_1 + \tilde{C}_1^T \tilde{C}_1 - Q, \\ \bar{A} &= A_1^T Q \tilde{B}_1, \\ M &= \gamma^2 I - \tilde{B}_1^T Q \tilde{B}_1 - \tilde{B}_2^T \tilde{C}_2^T \tilde{C}_2 \tilde{B}_2 > 0. \end{aligned}$$

Now, let a scalar $\alpha > 0$ and set

$$\begin{aligned} W &= -\tilde{C}_2^T \tilde{C}_2 - \alpha I, \\ Y &= \gamma^2 I - \tilde{B}_1^T Q \tilde{B}_1 - \tilde{B}_2^T W \tilde{B}_2. \end{aligned}$$

Then, it is easy to see that

$$Y = \gamma^2 I - \tilde{B}_1^T Q \tilde{B}_1 + \tilde{B}_2^T \tilde{C}_2^T \tilde{C}_2 \tilde{B}_2 + \alpha \tilde{B}_2^T \tilde{B}_2 \geq M > 0.$$

In the following, we will show that when α is sufficiently large, the following inequality holds:

$$\Delta + \bar{A} Y^{-1} \bar{A}^T + \left(\tilde{C}_1^T \tilde{C}_2 + \bar{A} Y^{-1} \tilde{B}_2^T W \right)$$
$$\times \tilde{B}_2 M^{-1} \tilde{B}_2^T \left(\tilde{C}_1^T \tilde{C}_2 + \bar{A} Y^{-1} \tilde{B}_2^T W \right)^T < 0. \tag{5.77}$$

To this end, we note that

$$\gamma^2 I - \tilde{B}_1^T Q \tilde{B}_1 > M > 0.$$

Therefore, there exists an invertible matrix Z such that

$$\gamma^2 I - \tilde{B}_1^T Q \tilde{B}_1 = Z^T Z. \qquad (5.78)$$

Observe that there exist orthogonal matrices U and V such that

$$\tilde{B}_2 Z^{-1} = U \begin{bmatrix} \Sigma & 0 \\ 0 & 0 \end{bmatrix} V, \qquad (5.79)$$

where $\Sigma > 0$ is a diagonal matrix. Write

$$U^T \tilde{C}_2^T \tilde{C}_2 U = \begin{bmatrix} H & * \\ * & * \end{bmatrix}, \qquad (5.80)$$

where the partition is compatible with that of $\tilde{B}_2 Z^{-1}$ in (5.79), and $*$ represents a matrix which will not be used in the following development. Now, by (5.78)–(5.80), we have

$$Y = Z^T V^T \begin{bmatrix} \Omega & 0 \\ 0 & I \end{bmatrix} V Z,$$

$$M^{-1} = Z^{-1} V^T \begin{bmatrix} (I - \Sigma H \Sigma)^{-1} & 0 \\ 0 & I \end{bmatrix} V Z^{-T},$$

where

$$\Omega = I + \Sigma H \Sigma + \alpha \Sigma^2. \qquad (5.81)$$

Then, by some algebraic manipulations, it can be verified that

$$\bar{A} Y^{-1} \bar{A}^T + \left(\tilde{C}_1^T \tilde{C}_2 + \bar{A} Y^{-1} \tilde{B}_2^T W \right)$$
$$\times \tilde{B}_2 M^{-1} \tilde{B}_2^T \left(\tilde{C}_1^T \tilde{C}_2 + \bar{A} Y^{-1} \tilde{B}_2^T W \right)^T$$
$$- \left(\tilde{C}_1^T \tilde{C}_2 \tilde{B}_2 - \bar{A} \right) M^{-1} \left(\tilde{C}_1^T \tilde{C}_2 \tilde{B}_2 - \bar{A} \right)^T$$
$$= \bar{A} Z^{-1} V^T \begin{bmatrix} S & 0 \\ 0 & 0 \end{bmatrix} V Z^{-T} \bar{A}^T + W_1 + W_1^T, \qquad (5.82)$$

where

$$S = \Omega^{-1} + \alpha^2 \Omega^{-1} \Sigma^2 (I - \Sigma H \Sigma)^{-1} \Sigma^2 \Omega^{-1} - (I - \Sigma H \Sigma)^{-1}$$
$$+ \Omega^{-1} \Sigma H \Sigma (I - \Sigma H \Sigma)^{-1} \Sigma H \Sigma \Omega^{-1}$$
$$+ \alpha \Omega^{-1} \Sigma^2 (I - \Sigma H \Sigma)^{-1} \Sigma H \Sigma \Omega^{-1}$$
$$+ \alpha \Omega^{-1} \Sigma H \Sigma (I - \Sigma H \Sigma)^{-1} \Sigma^2 \Omega^{-1}, \qquad (5.83)$$

$$W_1 = -\tilde{C}_1^T \tilde{C}_2 U \begin{bmatrix} \Upsilon & 0 \\ 0 & 0 \end{bmatrix} V Z^{-T} \bar{A}^T, \qquad (5.84)$$

and

$$\Upsilon = \Sigma (I - \Sigma H \Sigma)^{-1} \left(\alpha \Sigma^2 \Omega^{-1} - I + \Sigma H \Sigma \Omega^{-1} \right).$$

Considering (5.76), (5.81)–(5.84), it can be seen that when α is sufficiently large, the inequality in (5.77) holds. Now, by the matrix inversion lemma, we have

$$\left(-W - \tilde{C}_2^T \tilde{C}_2 - W \tilde{B}_2 Y^{-1} \tilde{B}_2^T W \right)^{-1} = \alpha^{-1} I + \alpha^{-2} W \tilde{B}_2 M_1^{-1} \tilde{B}_2^T W, \quad (5.85)$$

where

$$M_1 = M - \alpha^{-1} \tilde{B}_2^T \tilde{C}_2^T \tilde{C}_2 \tilde{C}_2^T \tilde{C}_2 \tilde{B}_2.$$

When α is sufficiently large, it follows from (5.77) and (5.85) that

$$\Delta + \bar{A} Y^{-1} \bar{A}^T + \left(\tilde{C}_1^T \tilde{C}_2 + \bar{A} Y^{-1} \tilde{B}_2^T W \right)$$
$$\times \left(W + \tilde{C}_2^T \tilde{C}_2 + W \tilde{B}_2 Y^{-1} \tilde{B}_2^T W \right)^{-1} \left(\tilde{C}_1^T \tilde{C}_2 + \bar{A} Y^{-1} \tilde{B}_2^T W \right)^T < 0,$$

which, by Schur complement, implies

$$\begin{bmatrix} \Delta + \bar{A} Y^{-1} \bar{A}^T & \tilde{C}_1^T \tilde{C}_2 + \bar{A} Y^{-1} \tilde{B}_2^T W \\ \left(\tilde{C}_1^T \tilde{C}_2 + \bar{A} Y^{-1} \tilde{B}_2^T W \right)^T & -W + \tilde{C}_2^T \tilde{C}_2 - W \tilde{B}_2 Y^{-1} \tilde{B}_2^T W \end{bmatrix} < 0. \quad (5.86)$$

Now, define

$$P = L_1^T \begin{bmatrix} Q & 0 \\ 0 & W \end{bmatrix} L_1. \quad (5.87)$$

Then, noting (5.86), we can see that $P = P^T$ given in (5.87) satisfies (5.68) and (5.69). This completes the proof. $\qquad\square$

Lemma 5.4. *Let*

$$\hat{E} = \begin{bmatrix} E & 0 \\ 0 & 0 \end{bmatrix}, \quad \hat{A} = \begin{bmatrix} A & 0 \\ 0 & I \end{bmatrix}, \quad \hat{B} = \begin{bmatrix} B \\ -D \end{bmatrix}, \quad \hat{C} = \begin{bmatrix} C & I \end{bmatrix}. \quad (5.88)$$

Then, there exists a matrix $\hat{P} = \hat{P}^T$ such that

$$\hat{E}^T \hat{P} \hat{E} \geq 0, \quad (5.89)$$

$$\begin{bmatrix} \hat{A}^T \hat{P} \hat{A} - \hat{E}^T \hat{P} \hat{E} + \hat{C}^T \hat{C} & \hat{A}^T \hat{P} \hat{B} \\ \hat{B}^T \hat{P} \hat{A} & \hat{B}^T \hat{P} \hat{B} - \gamma^2 I \end{bmatrix} < 0, \quad (5.90)$$

if and only if there exists a matrix $P = P^T$ satisfying the LMIs in (5.62) and (5.63).

Proof. (*Sufficiency*) Suppose that there exists a matrix $P = P^T$ satisfying the LMIs in (5.62) and (5.63). Then, it is easy to see that there exists a scalar $\delta > 0$ such that

$$\begin{bmatrix} A^T PA - E^T PE + C^T C & A^T PB + C^T D \\ B^T PA + D^T C & B^T PB + D^T D - \gamma^2 I \end{bmatrix} + \frac{1}{\delta} \begin{bmatrix} C^T \\ D^T \end{bmatrix} \begin{bmatrix} C & D \end{bmatrix} < 0,$$

which, by Schur complement, gives

$$\begin{bmatrix} A^T PA - E^T PE + C^T C & A^T PB + C^T D & C^T \\ B^T PA + D^T C & B^T PB + D^T D - \gamma^2 I & D^T \\ C & D & -\delta I \end{bmatrix} < 0.$$

Pre- and post-multiplying this inequality by

$$\begin{bmatrix} I & 0 & 0 \\ 0 & 0 & I \\ 0 & I & -D^T \end{bmatrix}$$

and its transpose, respectively, we obtain

$$\begin{bmatrix} A^T PA - E^T PE + C^T C & C^T & A^T PB \\ C & -\delta I & (\delta + 1) D \\ B^T PA & (\delta + 1) D^T & B^T PB - \gamma^2 I - (\delta + 1) D^T D \end{bmatrix} < 0 \tag{5.91}$$

Now, set

$$\hat{P} = \begin{bmatrix} P & 0 \\ 0 & -(\delta + 1) I \end{bmatrix}. \tag{5.92}$$

Then, by (5.91), it is easy to see that the matrix $\hat{P}$ given in (5.92) satisfies (5.89) and (5.90).

(*Necessity*) Suppose that there exists a matrix $\hat{P} = \hat{P}^T$ such that (5.89) and (5.90) hold. Then, write

$$\hat{P} = \begin{bmatrix} \hat{P}_1 & \hat{P}_2 \\ \hat{P}_2^T & \hat{P}_3 \end{bmatrix},$$

where the partition is compatible with that of $\hat{E}$ in (5.88). It is easy to see that (5.89) and (5.90) can be respectively rewritten as

$$\begin{bmatrix} E^T \hat{P}_1 E & 0 \\ 0 & 0 \end{bmatrix} \geq 0, \tag{5.93}$$

and

$$\begin{bmatrix} A^T \hat{P}_1 A - E^T \hat{P}_1 E + C^T C & A^T \hat{P}_2 + C^T & A^T \hat{P}_1 B - A^T \hat{P}_2 D \\ \hat{P}_2^T A + C & \hat{P}_3 + I & \hat{P}_2^T B - \hat{P}_3 D \\ B^T \hat{P}_1^T A - D^T \hat{P}_2^T A & B^T \hat{P}_2 - D^T \hat{P}_3 & \Psi \end{bmatrix} < 0, \tag{5.94}$$

where
$$\Psi = B^T \hat{P}_1 B - D^T \hat{P}_2^T B - B^T \hat{P}_2 D + D^T \hat{P}_3 D - \gamma^2 I.$$

Pre- and post-multiplying (5.94) by
$$\begin{bmatrix} I & 0 & 0 \\ 0 & D^T & I \end{bmatrix}$$

and its transpose, respectively, give
$$\begin{bmatrix} A^T \hat{P}_1 A - E^T \hat{P}_1 E + C^T C & A^T \hat{P}_1 B + C^T D \\ B^T \hat{P}_1^T A + D^T C & B^T \hat{P}_1 B + D^T D - \gamma^2 I \end{bmatrix} < 0. \tag{5.95}$$

Set $P = \hat{P}_1$. Then, it is easy to see that $P = P^T$ and (5.93) and (5.95) give (5.62) and (5.63), respectively. This completes the proof. $\square$

Proof of Theorem 5.5. Note that the discrete singular system in (5.58) and (5.59) is admissible and its transfer function satisfies (5.61) if and only if the discrete singular system
$$\hat{E}\hat{x}(k+1) = \hat{A}\hat{x}(k) + \hat{B}\omega(k),$$
$$z(k) = \hat{C}\hat{x}(k),$$

is admissible and its transfer function
$$\hat{G}(z) = \hat{C}\left(z\hat{E} - \hat{A}\right)^{-1}\hat{B},$$

satisfies
$$\|\hat{G}\|_\infty < \gamma.$$

This, by Lemma 5.3, is equivalent to the existence of a matrix $\hat{P} = \hat{P}^T$ satisfying (5.89) and (5.90) which, by Lemma 5.4, is equivalent to the existence of a matrix $P = P^T$ satisfying (5.62) and (5.63). This completes the proof. $\square$

It is noted that the bounded real lemma in Theorem 5.5 involves non-strict LMIs. We now provide a version of the bounded real lemma with a strict LMI in the following theorem.

Theorem 5.6. *Given a scalar $\gamma > 0$. The discrete singular system in (5.58) and (5.59) is admissible and its transfer function satisfies (5.61) if and only if there exist matrices $P > 0$ and Q such that the following LMI holds:*
$$\begin{bmatrix} A^T P A + C^T C - E^T P E & A^T P B + C^T D \\ B^T P A + D^T C & B^T P B + D^T D - \gamma^2 I \end{bmatrix}$$
$$+ \begin{bmatrix} A^T \\ B^T \end{bmatrix} S Q^T + Q S^T \begin{bmatrix} A^T \\ B^T \end{bmatrix}^T < 0, \tag{5.96}$$

where $S \in \mathbb{R}^{n \times (n-r)}$ is any matrix with full column rank and satisfies $E^T S = 0$.

Proof. (*Sufficiency*) Suppose that the LMI in (5.96) is satisfied. Under this condition, we first show that the discrete singular system in (5.58) and (5.59) is admissible. To this end, we write

$$Q = \begin{bmatrix} Q_1 \\ Q_2 \end{bmatrix},$$

where $Q_1 \in \mathbb{R}^{n \times (n-r)}$ and $Q_2 \in \mathbb{R}^{p \times (n-r)}$. Then, the LMI in (5.96) can be rewritten as

$$\begin{bmatrix} A^T P A + C^T C - E^T P E + A^T S Q_1^T + Q_1 S^T A \\ B^T P A + D^T C + Q_2 S^T A + B^T S Q_1^T \end{bmatrix}$$
$$\left. \begin{matrix} A^T P B + C^T D + A^T S Q_2^T + Q_1 S^T B \\ B^T P B + D^T D - \gamma^2 I + B^T S Q_2^T + Q_2 S^T B \end{matrix} \right] < 0, \qquad (5.97)$$

which implies

$$A^T P A - E^T P E + A^T S Q_1^T + Q_1 S^T A < 0.$$

Therefore, by Theorem 2.4, we have that system (5.58) is admissible. Next, we show that under the condition of the theorem, the discrete singular system in (5.58) and (5.59) satisfies the H_∞ performance in (5.61). For this purpose, we apply the Schur complement equivalence to (5.97) and obtain

$$A^T P A + C^T C - E^T P E + A^T S Q_1^T + Q_1 S^T A$$
$$+ \left(A^T P B + C^T D + A^T S Q_2^T + Q_1 S^T B \right)$$
$$\times U^{-1} \left(B^T P A + D^T C + Q_2 S^T A + B^T S Q_1^T \right) < 0, \qquad (5.98)$$

where
$$U = \gamma^2 I - D^T D - B^T P B - B^T S Q_2^T - Q_2 S^T B > 0. \qquad (5.99)$$

By (5.98), it is easy to see that there exists a matrix $W > 0$ such that

$$A^T P A + C^T C - E^T P E + A^T S Q_1^T + Q_1 S^T A$$
$$+ \left(A^T P B + C^T D + A^T S Q_2^T + Q_1 S^T B \right)$$
$$\times U^{-1} \left(B^T P A + D^T C + Q_2 S^T A + B^T S Q_1^T \right) + W < 0. \qquad (5.100)$$

Let
$$\Phi(j\theta) = e^{j\theta} E - A.$$

Then, considering the admissibility of system (5.58), it can be seen that $\Phi(j\theta)$ is nonsingular for all $\theta \in [0, 2\pi)$. Set

$$\Omega = \left(A^T P B + C^T D + A^T S Q_2^T + Q_1 S^T B \right)$$
$$\times U^{-1} \left(B^T P A + D^T C + Q_2 S^T A + B^T S Q_1^T \right) + W.$$

Then, pre- and post-multiplying (5.100) by $B^T \Phi(-j\theta)^{-T}$ and $\Phi(j\theta)^{-1} B$, respectively, yield

$$B^T \Phi(-j\theta)^{-T} \left[A^T P A + C^T C - E^T P E + A^T S Q_1^T + Q_1 S^T A \right]$$
$$\times \Phi(j\theta)^{-1} B + B^T \Phi(-j\theta)^{-T} \Omega \Phi(j\theta)^{-1} B \leq 0.$$

That is,

$$-B^T \Phi(-j\theta)^{-T} \left[A^T P A - E^T P E + A^T S Q_1^T + Q_1 S^T A \right] \Phi(j\theta)^{-1} B$$
$$\geq B^T \Phi(-j\theta)^{-T} \Omega \Phi(j\theta)^{-1} B + B^T \Phi(-j\theta)^{-T} C^T C \Phi(j\theta)^{-1} B. \qquad (5.101)$$

Now, by some simple algebraic manipulations, it can be verified that

$$B^T P B + B^T \Phi(-j\theta)^{-T} \left(A^T P + Q_1 S^T \right) B$$
$$+ B^T \left(A^T P + Q_1 S^T \right)^T \Phi(j\theta)^{-1} B + B^T \Phi(-j\theta)^{-T}$$
$$\times \left(A^T P A - E^T P E + A^T S Q_1^T + Q_1 S^T A \right) \Phi(j\theta)^{-1} B = 0, \qquad (5.102)$$

where the relationship $E^T S = 0$ is used. From (5.101) and (5.102), it is easy to show that for all $\theta \in [0, 2\pi)$,

$$B^T P B - B^T \Phi(-j\theta)^{-T} C^T C \Phi(j\theta)^{-1} B$$
$$\geq -B^T \Phi(-j\theta)^{-T} \left(A^T P + Q_1 S^T \right) B - B^T \left(A^T P + Q_1 S^T \right)^T \Phi(j\theta)^{-1} B$$
$$+ B^T \Phi(-j\theta)^{-T} \Omega \Phi(j\theta)^{-1} B. \qquad (5.103)$$

On the other hand, it can be deduced that for all $\theta \in [0, 2\pi)$,

$$\gamma^2 I - G(e^{-j\theta})^T G(e^{j\theta})$$
$$= \gamma^2 I - D^T D - B^T \Phi(-j\theta)^{-T} C^T C \Phi(j\theta)^{-1} B$$
$$- B^T \Phi(-j\theta)^{-T} C^T D - D^T C \Phi(j\theta)^{-1} B$$
$$= U + B^T P B - B^T \Phi(-j\theta)^{-T} C^T C \Phi(j\theta)^{-1} B$$
$$- B^T \Phi(-j\theta)^{-T} \left(A^T P B + C^T D + A^T S Q_2^T + Q_1 S^T B \right)$$
$$- \left(B^T P A + D^T C + Q_2 S^T A + B^T S Q_1^T \right) \Phi(j\theta)^{-1} B$$
$$+ B^T \Phi(-j\theta)^{-T} \left(A^T P + Q_1 S^T \right) B$$
$$+ B^T \left(A^T P + Q_1 S^T \right)^T \Phi(j\theta)^{-1} B \qquad (5.104)$$

where U is given in (5.99). Then, the equality in (5.104) together with the inequality in (5.103) implies that for all $\theta \in [0, 2\pi)$,

$$\gamma^2 I - G(e^{-j\theta})^T G(e^{j\theta})$$
$$\geq U - B^T \Phi(-j\theta)^{-T} \left(A^T P B + C^T D + A^T S Q_2^T + Q_1 S^T B \right)$$
$$- \left(B^T P A + D^T C + Q_2 S^T A + B^T S Q_1^T \right) \Phi(j\theta)^{-1} B$$
$$+ B^T \Phi(-j\theta)^{-T} \Omega \Phi(j\theta)^{-1} B.$$

By Lemma 4.6, it follows that

$$\gamma^2 I - G(e^{-j\theta})^T G(e^{j\theta})$$
$$\geq U - \left(B^T PA + D^T C + Q_2 S^T A + B^T SQ_1^T\right) \Omega^{-1}$$
$$\times \left(A^T PB + C^T D + A^T SQ_2^T + Q_1 S^T B\right). \tag{5.105}$$

Note that

$$\Omega - \left(A^T PB + C^T D + A^T SQ_2^T + Q_1 S^T B\right)$$
$$\times U^{-1} \left(B^T PA + D^T C + Q_2 S^T A + B^T SQ_1^T\right) = W > 0,$$

which, by Schur complement, implies

$$\begin{bmatrix} U & A^T PB + C^T D + A^T SQ_2^T + Q_1 S^T B \\ B^T PA + D^T C + Q_2 S^T A + B^T SQ_1^T & \Omega \end{bmatrix} > 0.$$

This, by Schur complement again, gives

$$U - \left(B^T PA + D^T C + Q_2 S^T A + B^T SQ_1^T\right)$$
$$\Omega^{-1} \left(A^T PB + C^T D + A^T SQ_2^T + Q_1 S^T B\right) > 0. \tag{5.106}$$

Therefore, from (5.105) and (5.106), we have

$$\gamma^2 I - G(e^{-j\theta})^T G(e^{j\theta}) > 0.$$

That is, the H_∞ performance in (5.61) is satisfied.

(*Necessity*) Suppose that the discrete singular system in (5.58) and (5.59) is admissible and its transfer function matrix satisfies (5.61). Then, there exist two nonsingular matrices $\mathcal{M}$ and $\mathcal{N}$ such that

$$E = \mathcal{M} \begin{bmatrix} I & 0 \\ 0 & 0 \end{bmatrix} \mathcal{N}, \quad A = \mathcal{M} \begin{bmatrix} A & 0 \\ 0 & I \end{bmatrix} \mathcal{N}.$$

Then S can be written as

$$S = \mathcal{M}^{-T} \begin{bmatrix} 0 \\ I \end{bmatrix} \mathcal{H},$$

where $\mathcal{H} \in \mathbb{R}^{(n-r) \times (n-r)}$ is any nonsingular matrix. Write

$$B = \mathcal{M} \begin{bmatrix} B_1 \\ B_2 \end{bmatrix}, \quad C = \begin{bmatrix} C_1 & C_2 \end{bmatrix} \mathcal{N},$$

where the partition is compatible with that of A. Noting

$$G(z) = C_1 \left(zI - A\right)^{-1} B_1 + D - C_2 B_2,$$

and using Lemma 5.2, we have that (5.61) ensures that there exists a matrix $\tilde{P} > 0$ such that

$$
\left[
\begin{array}{l}
\mathcal{A}^T \tilde{P} \mathcal{A} + C_1^T C_1 - \tilde{P} \\
B_1^T \tilde{P} \mathcal{A} + (D - C_2 B_2)^T C_1
\end{array}
\right.
$$
$$
\left.
\begin{array}{r}
\mathcal{A}^T \tilde{P} B_1 + C_1^T (D - C_2 B_2) \\
B_1^T \tilde{P} B_1 + (D - C_2 B_2)^T (D - C_2 B_2) - \gamma^2 I
\end{array}
\right] < 0.
\qquad (5.107)
$$

Set

$$
Q_1 = -C_1^T C_2, \quad Q_2 = -C_2^T C_2 / 2 - I,
$$
$$
Q_3 = -(D - C_2 B_2)^T C_2 + B_2^T Q_2.
$$

Then, by (5.107), it is easy to see that

$$
\left[
\begin{array}{ccc}
\mathcal{A}^T \tilde{P} \mathcal{A} + C_1^T C_1 - \tilde{P} & \mathcal{A}^T \tilde{P} B_1 + C_1^T (D - C_2 B_2) & \Upsilon_1 \\
B_1^T \tilde{P} \mathcal{A} + (D - C_2 B_2)^T C_1 & \Pi & \Upsilon_2 \\
\Upsilon_1^T & \Upsilon_2^T & \Upsilon_3
\end{array}
\right] < 0, \quad (5.108)
$$

where

$$
\Pi = B_1^T \tilde{P} B_1 + (D - C_2 B_2)^T (D - C_2 B_2) - \gamma^2 I,
$$
$$
\Upsilon_1 = C_1^T C_2 + Q_1,
$$
$$
\Upsilon_2 = Q_3 - B_2^T Q_2 + (D - C_2 B_2)^T C_2,
$$
$$
\Upsilon_3 = I + C_2^T C_2 + Q_2 + Q_2^T.
$$

Pre- and post-multiplying (5.108) by

$$
\left[
\begin{array}{ccc}
I & 0 & 0 \\
0 & 0 & I \\
0 & I & B_2^T
\end{array}
\right],
$$

and its transpose, respectively, result in

$$
\left[
\begin{array}{cc|c}
\mathcal{A}^T \tilde{P} \mathcal{A} + C_1^T C_1 - \tilde{P} & C_1^T C_2 + Q_1 & \Phi_1 \\
C_2^T C_1 + Q_1^T & I + C_2^T C_2 + Q_2 + Q_2^T & \Phi_2 \\
\hline
\Phi_1^T & \Phi_2^T & \Phi_3
\end{array}
\right] < 0
\qquad (5.109)
$$

where

$$
\Phi_1 = \mathcal{A}^T \tilde{P} B_1 + C_1^T D + Q_1 B_2,
$$
$$
\Phi_2 = B_2 + Q_3^T + Q_2 B_2 + C_2^T D,
$$
$$
\Phi_3 = B_1^T \tilde{P} B_1 + B_2^T B_2 + D^T D + B_2^T Q_3^T + Q_3 B_2 - \gamma^2 I.
$$

Let

$$\mathcal{P} = \mathcal{M}^{-T} \begin{bmatrix} \tilde{P} & 0 \\ 0 & I \end{bmatrix} \mathcal{M}, \quad \mathcal{Q} = \begin{bmatrix} \mathcal{N} & 0 \\ \hline 0 & I \end{bmatrix}^T \begin{bmatrix} Q_1 \\ Q_2 \\ Q_3 \end{bmatrix} \mathcal{H}^{-T}. \tag{5.110}$$

Then, pre- and post-multiplying (5.109) by

$$\begin{bmatrix} \mathcal{N} & 0 \\ \hline 0 & I \end{bmatrix}^T$$

and its transpose, we have

$$\begin{bmatrix} A^T \mathcal{P} A + C^T C - E^T \mathcal{P} E & A^T \mathcal{P} B + C^T D \\ B^T \mathcal{P} A + D^T C & B^T \mathcal{P} B + D^T D - \gamma^2 I \end{bmatrix}$$
$$+ \begin{bmatrix} A^T \\ B^T \end{bmatrix} S \mathcal{Q}^T + \mathcal{Q} S^T \begin{bmatrix} A^T \\ B^T \end{bmatrix}^T < 0. \tag{5.111}$$

That is, the matrices $\mathcal{P}$ and $\mathcal{Q}$ given in (5.110) satisfy (5.111). This completes the proof. $\qquad\square$

Now, we consider the following state feedback controller

$$u(k) = Kx(k), \quad K \in \mathbb{R}^{m \times n}. \tag{5.112}$$

Applying this to (5.56) and (5.57), we obtain the closed-loop system as

$$Ex(k+1) = (A + B_1 K) x(k) + Bw(k), \tag{5.113}$$
$$z(k) = Cx(k) + Dw(k), \tag{5.114}$$

then we have the following H_∞ control result.

Theorem 5.7. *Given a scalar $\gamma > 0$ and the discrete singular system in (5.56) and (5.57). There exists a state feedback controller (5.112) such that the closed-loop system in (5.113) and (5.114) is admissible and its transfer function*

$$G_{dK}(z) = C \left[zI - (A + B_1 K) \right]^{-1} B + D,$$

satisfies

$$\|G_{dK}\|_\infty < \gamma, \tag{5.115}$$

if and only if there exist a scalar $\delta > 0$, matrices $P > 0$, Q_1 and Q_2 such that

$$\tilde{X} = \gamma^2 I - \left[B^T PB + D^T D + Q_2 S^T B + B^T S Q_2^T \right] > 0, \tag{5.116}$$

and

$$\tilde{Y}_1 + \tilde{Y}_2 \tilde{X}^{-1} \tilde{Y}_2^T - \left(\tilde{\Psi}_1^T + \tilde{Y}_2 \tilde{X}^{-1} \tilde{\Psi}_2^T \right) \tilde{\Pi}^{-1} \left(\tilde{\Psi}_1 + \tilde{\Psi}_2 \tilde{X}^{-1} \tilde{Y}_2^T \right) < 0, \tag{5.117}$$

where $S \in \mathbb{R}^{n \times (n-r)}$ is any matrix with full column rank and satisfies $E^T S = 0$, and

$$\tilde{Y}_1 = C^T C - E^T PE + Q_1 S^T A + A^T S Q_1^T + A^T PA,$$
$$\tilde{Y}_2 = C^T D + Q_1 S^T B + A^T \left(PB + SQ_2^T\right),$$
$$\tilde{\Psi}_1 = B_1^T \left(PA + SQ_1^T\right),$$
$$\tilde{\Psi}_2 = B_1^T \left(PB + SQ_2^T\right),$$
$$\tilde{\Pi} = B_1^T PB_1 + \tilde{\Psi}_2 \tilde{X}^{-1} \tilde{\Psi}_2^T + \delta I.$$

In this case, a desired state feedback control law can be chosen by

$$u(k) = -\tilde{\Pi}^{-1} \left(\tilde{\Psi}_1 + \tilde{\Psi}_2 \tilde{X}^{-1} \tilde{Y}_2^T\right) x(k). \tag{5.118}$$

Proof. By Theorem 5.6, it is easy to see that there exists a state feedback controller in (5.112) such that the closed-loop system in (5.113) and (5.114) is admissible and its transfer function satisfies (5.115) if and only if there exist matrices $P > 0$ and Q such that

$$\begin{bmatrix} A_c^T PA_c + C^T C - E^T PE & A_c^T PB + C^T D \\ B^T PA + D^T C & B^T PB + D^T D - \gamma^2 I \end{bmatrix}$$
$$+ \begin{bmatrix} A_c^T \\ B^T \end{bmatrix} SQ^T + QS^T \begin{bmatrix} A_c^T \\ B^T \end{bmatrix}^T < 0, \tag{5.119}$$

where

$$A_c = A + B_1 K.$$

Write

$$Q = \begin{bmatrix} Q_1 \\ Q_2 \end{bmatrix}, \tag{5.120}$$

where the partition is compatible with other matrices. Then, by Schur complement and the partition in (5.120), the inequality in (5.119) can be rewritten as

$$\begin{bmatrix} C^T C - E^T PE + Q_1 S^T A_c + A_c^T S Q_1^T \\ \left(B^T P + Q_2 S^T\right) A_c + D^T C + B^T S Q_1^T \\ PA_c \end{bmatrix}$$
$$\begin{matrix} A_c^T \left(PB + SQ_2^T\right) + C^T D + Q_1 S^T B & A_c^T P \\ B^T PB + D^T D - \gamma^2 I + Q_2 S^T B + B^T SQ_2 & 0 \\ 0 & -P \end{matrix} \Bigg] < 0,$$

which, by Schur complement, is equivalent to $\tilde{X} > 0$, and

$$\begin{bmatrix} \tilde{Y}_1 + \tilde{\Psi}_1^T K + K^T \tilde{\Psi}_1 + K^T B_1^T PB_1 K & \tilde{Y}_2 + K^T \tilde{\Psi}_2 \\ \tilde{Y}_2^T + \tilde{\Psi}_2^T K & -\tilde{X} \end{bmatrix} < 0. \tag{5.121}$$

Using Schur complement again, we have that (5.121) is equivalent to

$$\tilde{Y}_1 + \tilde{Y}_2 \tilde{X}^{-1} \tilde{Y}_2^T + \left(\tilde{\Psi}_1^T + \tilde{Y}_2 \tilde{X}^{-1} \tilde{\Psi}_2^T \right) K + K^T \left(\tilde{\Psi}_1 + \tilde{\Psi}_2 \tilde{X}^{-1} \tilde{Y}_2^T \right)$$

$$+ K^T \left(B_1^T P B_1 + \tilde{\Psi}_2 \tilde{X}^{-1} \tilde{\Psi}_2^T \right) K < 0.$$

This inequality is satisfied if and only if there exists a scalar $\delta > 0$ such that

$$\tilde{Y}_1 + \tilde{Y}_2 \tilde{X}^{-1} \tilde{Y}_2^T + \left(\tilde{\Psi}_1^T + \tilde{Y}_2 \tilde{X}^{-1} \tilde{\Psi}_2^T \right) K$$

$$+ K^T \left(\tilde{\Psi}_1 + \tilde{\Psi}_2 \tilde{X}^{-1} \tilde{Y}_2^T \right) + K^T \tilde{\Pi} K < 0.$$

That is,

$$\tilde{Y}_1 + \tilde{Y}_2 \tilde{X}^{-1} \tilde{Y}_2^T - \left(\tilde{\Psi}_1^T + \tilde{Y}_2 \tilde{X}^{-1} \tilde{\Psi}_2^T \right) \tilde{\Pi}^{-1} \left(\tilde{\Psi}_1 + \tilde{\Psi}_2 \tilde{X}^{-1} \tilde{Y}_2^T \right)$$

$$+ \left[K^T + \left(\tilde{\Psi}_1^T + \tilde{Y}_2 \tilde{X}^{-1} \tilde{\Psi}_2^T \right) \tilde{\Pi}^{-1} \right]$$

$$\times \tilde{\Pi} \left[K + \tilde{\Pi}^{-1} \left(\tilde{\Psi}_1 + \tilde{\Psi}_2 \tilde{X}^{-1} \tilde{Y}_2^T \right) \right] < 0. \tag{5.122}$$

Then, it is easy to see that there exists a matrix K such that (5.122) holds if and only if (5.117) holds and, in this case, a suitable K can be chosen as in (5.118). This completes the proof. $\square$

5.4 Robust H_∞ Control

In this section, the problem of robust H_∞ control for singular systems is addressed. The notion of generalized quadratic stabilizability with an H_∞-norm bound is proposed and state feedback controllers are designed for uncertain continuous and discrete singular systems, respectively.

5.4.1 Continuous Systems

Consider an uncertain linear continuous singular system described by

$$E\dot{x}(t) = (A + \Delta A)\, x(t) + (B_1 + \Delta B_1)u(t) + Bw(t), \tag{5.123}$$
$$z(t) = Cx(t) + B_2 u(t), \tag{5.124}$$

where ΔA and ΔB_1 are unknown time-invariant matrices representing norm-bounded parameter uncertainties, which are assumed to be of the form:

$$\begin{bmatrix} \Delta A \ \Delta B_1 \end{bmatrix} = MF(\sigma) \begin{bmatrix} N_1 \ N_2 \end{bmatrix} \tag{5.125}$$

where M, N_1 and N_2 are known real constant matrices with appropriate dimensions. The uncertain matrix $F(\sigma)$ satisfies

$$F(\sigma)^T F(\sigma) \leq I, \tag{5.126}$$

where $\sigma \in \Theta$, and Θ is a compact set in $\mathbb{R}$. Furthermore, it is assumed that given any matrix $F : F^T F \leq I$, there exists a $\sigma \in \Theta$ such that $F = F(\sigma)$.

Definition 5.1. *Let the constant scalar $\gamma > 0$ be given. The uncertain continuous singular system in (5.123) and (5.124) is said to be robustly stabilizable with an H_∞-norm γ if there exists a linear state feedback control law*

$$u(t) = Kx(t), \quad K \in \mathbb{R}^{m \times n}, \tag{5.127}$$

such that the closed-loop system is admissible and its transfer function

$$G_{cK}(s) = C_c \left[sI - (A_c + \Delta A_{cK}) \right]^{-1} B,$$

satisfies

$$\|G_{cK}\|_\infty < \gamma, \tag{5.128}$$

for all allowable uncertainties, where

$$A_c = A + B_1 K, \tag{5.129}$$
$$C_c = C + B_2 K, \tag{5.130}$$
$$\Delta A_{cK} = \Delta A + \Delta B_1 K. \tag{5.131}$$

Thus, the robust H_∞ control problem to be investigated is formulated as follows: given a constant scalar $\gamma > 0$, develop a condition to ensure that the singular system in (5.123) and (5.124) is robustly stabilizable with an H_∞-norm γ.

Note that in the context of state-space systems, the concept of quadratic stabilizability with an H_∞-norm bound has played an important role in solving the robust H_∞ control problem. Considering this, we introduce the following definition.

Definition 5.2. *Let the constant scalar $\gamma > 0$ be given. The uncertain continuous singular system in (5.123) and (5.124) is said to be generalized quadratically stabilizable with an H_∞-norm γ if there exists a linear state feedback control law (5.127) and matrices $P > 0$ and Q such that for all allowable uncertainties, the following LMI holds:*

$$\begin{bmatrix} \Lambda_{cK} + BB^T & \left(PE^T + SQ\right)^T C_c^T \\ C_c \left(PE^T + SQ\right) & -\gamma^2 I \end{bmatrix} < 0, \qquad (5.132)$$

where

$$\Lambda_{cK} = \left(PE^T + SQ\right)^T \left(A_c + \Delta A_{cK}\right)^T + \left(A_c + \Delta A_{cK}\right) \left(PE^T + SQ\right),$$

the matrices A_c, C_c and ΔA_{cK} are given in (5.129)–(5.131), respectively, and $S \in \mathbb{R}^{n \times (n-r)}$ is any matrix with full column rank and satisfies $ES = 0$.

The following lemma shows that generalized quadratic stabilizability with an H_∞-norm bound implies robust stabilizability with an H_∞-norm bound.

Lemma 5.5. *If the uncertain continuous singular system in (5.123) and (5.124) is generalized quadratically stabilizable with an H_∞-norm γ, then this system is robustly stabilizable with an H_∞-norm γ.*

Proof. Suppose that the uncertain continuous singular system in (5.123) and (5.124) is generalized quadratically stabilizable with an H_∞-norm γ. Then, by Definition 5.2, we have that there exists a linear state feedback control law (5.127) and matrices $P > 0$ and Q such that (5.132) is satisfied. This, by Corollary 5.2, implies that the closed-loop system is admissible and the H_∞ performance in (5.128) is satisfied. Therefore, by Definition 5.1, it is easy to see that the uncertain continuous singular system in (5.123) and (5.124) is robustly stabilizable with an H_∞-norm γ. □

Taking into account of Lemma 5.5, our attention will be focused on the development of a condition for generalized quadratic stabilizability with an H_∞-norm bound. A necessary and sufficient condition is presented in the following theorem. The proof can be carried out along a similar line as in the proof of Theorem 4.7, and is thus omitted.

Theorem 5.8. *Given a constant scalar $\gamma > 0$. The uncertain continuous singular system in (5.123) and (5.124) is generalized quadratically stabilizable with an H_∞-norm $\gamma > 0$ if and only if there exist matrices $P > 0$, Q, Y and a scalar $\epsilon > 0$ such that*

$$\begin{bmatrix} \varXi\left(P,Q\right) + \epsilon M M^T & \left(PE^T + SQ\right)^T C^T + Y^T B_2^T \\ C\left(PE^T + SQ\right) + B_2 Y & -\gamma^2 I \\ N_1 \Omega\left(P,Q\right) + N_2 Y & 0 \end{bmatrix}$$
$$\begin{matrix} \Omega\left(P,Q\right)^T N_1^T + Y^T N_2^T \\ 0 \\ -\epsilon I \end{matrix} \Bigg] < 0, \qquad (5.133)$$

where

$$\Xi(P,Q) = \Omega(P,Q)^T A^T + A\Omega(P,Q) + B_1 Y + Y^T B_1^T,$$
$$\Omega(P,Q) = PE^T + SQ,$$

and $S \in \mathbb{R}^{n \times (n-r)}$ is any matrix with full column rank and satisfies $ES = 0$. In this case, we can assume that the matrix $\Omega(P,Q)$ is nonsingular (if this is not the case, then we can choose some $\theta \in (0,1)$ such that $\hat{\Omega}(P,Q) = \Omega(P,Q) + \theta \tilde{P}$ is nonsingular and satisfies (5.133), in which $\tilde{P}$ is any nonsingular matrix satisfying $E\tilde{P} = \tilde{P}^T E^T \geq 0$), then a desired robustly stabilizing state feedback controller can be chosen as

$$u(t) = Y\Omega(P,Q)^{-1} x(t).$$

To illustrate the applicability of Theorem 5.8, we provide the following example.

Example 5.1. Consider an uncertain continuous singular system in (5.123) and (5.124) with parameters as

$$E = \begin{bmatrix} -10.5 & -7 & 7 \\ -18 & 7.5 & 5.5 \\ 7.5 & -4 & -2 \end{bmatrix}, \quad A = \begin{bmatrix} 31.5 & 0 & -14 \\ 15 & 13 & -11 \\ -4.5 & -6 & 4 \end{bmatrix},$$

$$B_1 = \begin{bmatrix} 1.9 & -0.3 \\ -1 & -1.8 \\ 0.3 & 1.3 \end{bmatrix}, \quad B = \begin{bmatrix} 2.7 \\ 2.6 \\ -1.3 \end{bmatrix},$$

$$B_2 = \begin{bmatrix} -1 & 1 \end{bmatrix}, \quad C = \begin{bmatrix} 5.3 & -2 & -2.8 \end{bmatrix},$$

$$M = \begin{bmatrix} -0.1 & 0.1 \\ -0.45 & 0.85 \\ 0.2 & -0.2 \end{bmatrix}, \quad N_1 = \begin{bmatrix} 0.25 & 1.3 & -0.1 \\ 1.1 & 0 & -0.6 \end{bmatrix},$$

$$N_2 = \begin{bmatrix} 0 & 1 \\ -1 & 0.5 \end{bmatrix}.$$

It can be verified that the nominal system is not regular, unstable and impulsive. The purpose is to design a state feedback controller such that the closed-loop system is admissible and the transfer function of the closed-loop system satisfies a prescribed H_∞-norm bound γ. In this example, we suppose the required H_∞-norm bound is specified as $\gamma = 0.5$. To solve the robust H_∞ control problem, we note that rank$(E) = 2$. Therefore, we can choose

$$S = \begin{bmatrix} 0.8 \\ 0.6 \\ 1.8 \end{bmatrix},$$

which is with full column rank and satisfies $ES = 0$. It is found that the LMI in (5.133) is feasible and a set of solutions is obtained as follows:

$$P = \begin{bmatrix} 0.8650 & 0.6389 & 1.9220 \\ 0.6389 & 0.5223 & 1.4315 \\ 1.9220 & 1.4315 & 4.3373 \end{bmatrix}, \quad Y = \begin{bmatrix} 0.2030 & 4.4424 & -1.5481 \\ -1.8838 & 1.1525 & -0.8939 \end{bmatrix},$$

$$Q = \begin{bmatrix} -1.1699 & -2.4850 & 0.7100 \end{bmatrix}, \quad \epsilon = 5.3860.$$

In this case, by Theorem 5.8, a desired state feedback controller can be constructed as

$$u(t) = \begin{bmatrix} 0.0347 & 3.2954 & -1.8932 \\ -5.4909 & 5.4209 & 0.9811 \end{bmatrix} x(t).$$

$\diamond$

In the case when $E = I$, that is, the uncertain continuous singular system in (5.123) and (5.124) reduces to the following uncertain state-space system:

$$\dot{x}(t) = (A + \Delta A)\,x(t) + (B_1 + \Delta B_1)u(t) + B\omega(t), \qquad (5.134)$$
$$z(t) = Cx(t) + B_2 u(t). \qquad (5.135)$$

By Theorem 5.8, it is easy to have the following result.

Corollary 5.3. *Given a constant scalar $\gamma > 0$. The uncertain continuous state-space system in (5.134) and (5.135) is quadratically stabilizable with an H_∞ norm $\gamma > 0$ if and only if there exist matrices $P > 0$, Y and a scalar $\epsilon > 0$ such that*

$$\begin{bmatrix} \Xi + \epsilon MM^T & PC^T + Y^T B_2^T & PN_1^T + Y^T N_2^T \\ CP + B_2 Y & -\gamma^2 I & 0 \\ N_1 P + N_2 Y & 0 & -\epsilon I \end{bmatrix} < 0,$$

where

$$\Xi = PA^T + AP + BY + Y^T B^T.$$

In this case, a desired robustly stabilizing state feedback controller can be chosen as

$$u(t) = YP^{-1}x(t).$$

5.4.2 Discrete Systems

Now, we consider the robust H_∞ control problem for uncertain discrete singular systems via state feedback controllers. The uncertain discrete singular system to be considered is described by

$$Ex(k+1) = (A + \Delta A)x(k) + (B_1 + \Delta B_1)u(k) + B\omega(k), \quad (5.136)$$
$$z(k) = Cx(k) + D\omega(k), \quad (5.137)$$

where ΔA and ΔB_1 are time-invariant matrices representing norm-bounded parameter uncertainties, and are assumed to be of the form (5.125) and (5.126).

Definition 5.3. *Let the constant scalar $\gamma > 0$ be given. The uncertain discrete singular system in (5.136) and (5.137) is said to be robustly stabilizable with an H_∞-norm bound γ if there exists a linear state feedback control law*

$$u(k) = Kx(k), \quad K \in \mathbb{R}^{m \times n}, \quad (5.138)$$

such that the closed-loop system is admissible and its transfer function

$$G_{dK}(z) = C\left[zI - (A_d + \Delta A_{dK})\right]^{-1} B + D,$$

satisfies

$$\|G_{dK}\|_\infty < \gamma,$$

for all admissible uncertainties, where

$$A_d = A + B_1 K, \quad \Delta A_{dK} = \Delta A + \Delta B_1 K. \quad (5.139)$$

The problem to be addressed is formulated as follows: given a constant scalar $\gamma > 0$, develop a condition to ensure robust stabilizability with an H_∞-norm bound γ for the uncertain discrete-time singular system in (5.136) and (5.137).

Similar to the continuous case, we introduce the following definition.

Definition 5.4. *Let the constant scalar $\gamma > 0$ be given. The unforced discrete singular system in (5.136) and (5.137) is said to be generalized quadratically stabilizable with an H_∞-norm bound γ if there exist a linear state feedback control law (5.138) and matrices $P > 0$ and Q such that for all allowable uncertainties, the following LMI holds:*

$$\begin{bmatrix} (A_d + \Delta A_{dK})^T P (A_d + \Delta A_{dK}) + C^T C - E^T PE \\ B^T P (A_d + \Delta A_{dK}) + D^T C \end{bmatrix}$$
$$\begin{bmatrix} (A_d + \Delta A_{dK})^T PB + C^T D \\ B^T PB + D^T D - \gamma^2 I \end{bmatrix}$$
$$+ \begin{bmatrix} (A_d + \Delta A_{dK})^T \\ B^T \end{bmatrix} SQ^T + QS^T \begin{bmatrix} (A_d + \Delta A_{dK})^T \\ B^T \end{bmatrix}^T < 0, \quad (5.140)$$

where $S \in \mathbb{R}^{n \times (n-r)}$ is any matrix with full column rank and satisfies $E^T S = 0$.

The following lemma shows that generalized quadratic stabilizability with an H_∞-norm bound implies robust stabilizability with an H_∞-norm bound. The proof can be easily carried out by following a similar line as in the proof of Lemma 5.5, and is thus omitted.

Lemma 5.6. *Given a constant scalar $\gamma > 0$. The uncertain discrete singular system in (5.136) and (5.137) is robustly stabilizable with an H_∞-norm bound γ if the uncertain discrete singular system in (5.136) and (5.137) is generalized quadratically stabilizable with an H_∞-norm bound γ.*

In view of this, in the following, attention will be focused on the development of a condition for generalized quadratic stabilizability with an H_∞-norm bound; such a condition is given in the following theorem.

Theorem 5.9. *The uncertain discrete singular system in (5.136) and (5.137) is generalized quadratically stabilizable with an H_∞-norm bound γ if and only if there exist scalars $\epsilon > 0$, $\delta > 0$, matrices $P > 0$, Q_1 and Q_2 such that*

$$Z = P^{-1} - \epsilon^{-1} M M^T > 0, \tag{5.141}$$

$$\begin{aligned}
X = \gamma^2 I &- \big[B^T P B + D^T D + Q_2 S^T B + B^T S Q_2^T \\
&+ \epsilon^{-1} \left(B^T P + Q_2 S^T \right) M \left(I + \epsilon^{-1} M^T Z^{-1} M \right) \\
&\times M^T \left(B^T P + Q_2 S^T \right)^T \big] > 0,
\end{aligned} \tag{5.142}$$

and

$$Y_1 + Y_2 X^{-1} Y_2^T - \left(\Psi_1^T + Y_2 X^{-1} \Psi_2^T \right) \Pi^{-1} \left(\Psi_1 + \Psi_2 X^{-1} Y_2^T \right) < 0, \tag{5.143}$$

where the matrix $S \in \mathbb{R}^{n \times (n-r)}$ is any matrix with full column rank and satisfies $E^T S = 0$, and

$$\begin{aligned}
Y_1 &= C^T C - E^T P E + \epsilon N_1^T N_1 + Q_1 S^T A + A^T S Q_1^T \\
&\quad + \epsilon^{-1} Q_1 S^T M M^T S Q_1^T + \left(A^T + \epsilon^{-1} Q_1 S^T M M^T \right) \\
&\quad \times Z^{-1} \left(A + \epsilon^{-1} M M^T S Q_1^T \right), \\
Y_2 &= C^T D + Q_1 S^T B + \left(A^T + \epsilon^{-1} Q_1 S^T M M^T \right) \\
&\quad \times \left(I + \epsilon^{-1} Z^{-1} M M^T \right) \left(P B + S Q_2^T \right), \\
\Psi_1 &= \epsilon N_2^T N_1 + B_1^T Z^{-1} A + B_1^T \left(I + \epsilon^{-1} Z^{-1} M M^T \right) S Q_1^T, \\
\Psi_2 &= B_1^T \left(I + \epsilon^{-1} Z^{-1} M M^T \right) \left(P B + S Q_2^T \right), \\
\Pi &= \epsilon N_2^T N_2 + B_1^T Z^{-1} B_1 + \Psi_2 X^{-1} \Psi_2^T + \delta I.
\end{aligned}$$

In this case, a desired robustly stabilizing state feedback control law can be chosen by

$$u(k) = -\Pi^{-1} \left(\Psi_1 + \Psi_2 X^{-1} Y_2^T \right) x(k). \tag{5.144}$$

Proof. By Definition 5.4, it follows that the uncertain discrete singular system in (5.136) and (5.137) is generalized quadratically stabilizable with an H_∞-norm bound γ if and only if there exist a linear state feedback control law in (5.138) and matrices $P > 0$ and Q such that (5.140) is satisfied. Following a similar line as in the proof of Theorem 4.4, we can deduce that (5.140) is equivalent to the existence of a scalar $\epsilon > 0$, matrices $P > 0$, Q and $K \in \mathbb{R}^{m \times n}$ such that

$$\begin{bmatrix} \hat{\Theta}_1 + \check{\Theta}_1 & \hat{\Theta}_2 & \hat{\Theta}_3 M \\ \hat{\Theta}_2^T & -P & PM \\ M^T \hat{\Theta}_3^T & M^T P & -\epsilon I \end{bmatrix} < 0, \tag{5.145}$$

where

$$\hat{\Theta}_3 = \begin{bmatrix} 0 \\ B^T P \end{bmatrix} + QS^T,$$

$$\hat{\Theta}_1 = \begin{bmatrix} C^T C - E^T PE & A_{dK}^T PB + C^T D \\ B^T PA_{dK} + D^T C & B^T PB + D^T D - \gamma^2 I \end{bmatrix}$$

$$+ \begin{bmatrix} A_{dK}^T \\ B^T \end{bmatrix} SQ^T + QS^T \begin{bmatrix} A_{dK}^T \\ B^T \end{bmatrix}^T,$$

$$\check{\Theta}_1 = \begin{bmatrix} \epsilon N_{dK}^T N_{dK} & 0 \\ 0 & 0 \end{bmatrix}, \quad \hat{\Theta}_2 = \begin{bmatrix} A_{dK}^T P \\ 0 \end{bmatrix},$$

$$N_{dK} = N_1 + N_2 K.$$

Write

$$Q = \begin{bmatrix} Q_1 \\ Q_2 \end{bmatrix},$$

where the partition is compatible with other matrices. Then, (5.145) can be rewritten as

$$\begin{bmatrix} W_1 & A_c^T \left(PB + SQ_2^T \right) \\ & +C^T D + Q_1 S^T B \\ \left(B^T P + Q_2 S^T \right) A_c & W_2 \\ +D^T C + B^T SQ_1^T & \\ PA_c & 0 \\ M^T SQ_1^T & M^T \left(PB + SQ_2^T \right) \\ A_c^T P & Q_1 S^T M \\ 0 & \left(B^T P + Q_2 S^T \right) M \\ -P & PM \\ M^T P & -\epsilon I \end{bmatrix} < 0, \tag{5.146}$$

where

$$W_1 = C^T C - E^T PE + \epsilon N_c^T N_c + Q_1 S^T A_c + A_c^T SQ_1^T,$$
$$W_2 = B^T PB + D^T D - \gamma^2 I + Q_2 S^T B + B^T SQ_2.$$

By Schur complement, it can be seen that (5.146) is equivalent to $Z > 0$, $X > 0$, and

$$\begin{bmatrix} Y_1 + \Psi_1^T K + K^T \Psi_1 + K^T \left(\epsilon N_2^T N_2 + B_1^T Z^{-1} B_1 \right) K \ Y_2 + K^T \Psi_2 \\ Y_2^T + \Psi_2^T K & -X \end{bmatrix} < 0.$$
$$(5.147)$$

Using Schur complement again, we have that (5.147) is equivalent to

$$Y_1 + Y_2 X^{-1} Y_2^T + \left(\Psi_1^T + Y_2 X^{-1} \Psi_2^T \right) K + K^T \left(\Psi_1 + \Psi_2 X^{-1} Y_2^T \right)$$
$$+ K^T \left(\epsilon N_2^T N_2 + B_1^T Z^{-1} B_1 + \Psi_2 X^{-1} \Psi_2^T \right) K < 0.$$

This inequality is satisfied if and only if there exists a scalar $\delta > 0$ such that

$$Y_1 + Y_2 X^{-1} Y_2^T + \left(\Psi_1^T + Y_2 X^{-1} \Psi_2^T \right) K$$
$$+ K^T \left(\Psi_1 + \Psi_2 X^{-1} Y_2^T \right) + K^T \Pi K < 0.$$

That is,

$$Y_1 + Y_2 X^{-1} Y_2^T - \left(\Psi_1^T + Y_2 X^{-1} \Psi_2^T \right) \Pi^{-1} \left(\Psi_1 + \Psi_2 X^{-1} Y_2^T \right)$$
$$+ \left[K^T + \left(\Psi_1^T + Y_2 X^{-1} \Psi_2^T \right) \Pi^{-1} \right]$$
$$\times \Pi \left[K + \Pi^{-1} \left(\Psi_1 + \Psi_2 X^{-1} Y_2^T \right) \right] < 0. \tag{5.148}$$

Then, it is easy to see that there exists a matrix K such that (5.148) holds if and only if (5.143) holds and, in this case, a suitable K can be chosen as in (5.144). This completes the proof. $\qquad \square$

Now we provide an example to illustrate the application of the proposed method in Theorem 5.9.

Example 5.2. Consider an uncertain discrete-time singular system in (5.136) and (5.137) with the following parameters:

$$E = \begin{bmatrix} 2.6 & 0.3 & 1.3 \\ 0.5 & 0 & 0.25 \\ 1 & 1.5 & 0.5 \end{bmatrix}, \quad A = \begin{bmatrix} 6 & 0.85 & 3 \\ 1.2 & 0.1 & 0.6 \\ 1.2 & 1.85 & 0.6 \end{bmatrix},$$

$$B = \begin{bmatrix} -0.1 \\ 0.6 \\ -1.4 \end{bmatrix}, \quad B_1 = \begin{bmatrix} 1.3 & 4 & 3 \\ -1 & -1.5 & -0.5 \\ 3.5 & 2 & 0 \end{bmatrix},$$

$$M = \begin{bmatrix} 1.29 \\ 0.05 \\ -0.15 \end{bmatrix}, \quad C = \begin{bmatrix} -0.05 & -0.4 & -0.9 \end{bmatrix},$$

$$D = 0.2, \quad N_1 = \begin{bmatrix} 0.25 & 0.1 & 0.3 \end{bmatrix}, \quad N_2 = \begin{bmatrix} 0.3 & 0.1 & 0.2 \end{bmatrix}.$$

It can be seen that the nominal discrete-singular system is not regular, non-causal, and unstable. The purpose is to design a state feedback controller such that, for all allowable uncertainties, the closed-loop system is admissible and the transfer function from exogenous disturbance to the controlled output satisfies a prescribed H_∞ norm bound constraint. In this example, we suppose the required H_∞ norm bound is specified as $\gamma = 1.8$. To solve the problem, we choose

$$S = \begin{bmatrix} 0.1852 & -0.8889 & -0.0370 \end{bmatrix}^T,$$

which is with full column rank and satisfies $E^T S = 0$. Now, it can be verified that

$$P = \begin{bmatrix} 1.2765 & 0.1280 & 0.1974 \\ 0.1280 & 3.5830 & 1.3571 \\ 0.1974 & 1.3571 & 0.7589 \end{bmatrix}, \quad Q_1 = \begin{bmatrix} 0.7 \\ -0.4 \\ -1.4 \end{bmatrix},$$

$$Q_2 = 0.1, \quad \delta = 0.02, \quad \epsilon = 2.2,$$

satisfy the matrix inequalities in (5.141)–(5.143). Therefore, by Theorem 5.9, a desired state feedback control law can be chosen as

$$u(k) = \begin{bmatrix} -2.7740 & -1.2628 & -1.7340 \\ 3.2596 & 1.3239 & 3.1618 \\ -5.1176 & -1.4867 & -4.4273 \end{bmatrix} x(k).$$

$\diamond$

In the case when $E = I$, that is, the uncertain discrete singular system in (5.136) and (5.137) reduces to the following uncertain state-space system:

$$x(k+1) = (A + \Delta A)x(k) + (B_1 + \Delta B_1)u(k) + Bw(k), \qquad (5.149)$$
$$z(k) = Cx(k) + Dw(k). \qquad (5.150)$$

By Theorem 5.9, it is easy to have the following result.

Corollary 5.4. *Given a constant scalar $\gamma > 0$. The uncertain discrete state-space system in (5.149) and (5.150) is quadratically stabilizable with an H_∞-norm γ if and only if there exist matrices $P > 0$, Y and a scalar $\epsilon > 0$ such that*

$$\begin{bmatrix} -P & 0 & PC^T & PA^T + Y^T B_1^T & PN_1^T + Y^T N_2^T \\ 0 & -I & D^T & B^T & 0 \\ CP & D & -\gamma^2 I & 0 & 0 \\ AP + B_1 Y & B & 0 & \epsilon M M^T - P & 0 \\ N_1 P + N_2 Y & 0 & 0 & 0 & -\epsilon I \end{bmatrix} < 0,$$

In this case, a desired robustly stabilizing state feedback controller can be chosen as

$$u(k) = Y P^{-1} x(k).$$

5.5 Conclusion

This chapter has studied the problem of robust H_∞ control for uncertain singular systems. Versions of the bounded real lemma for both continuous and discrete singular systems have been presented. Based on these, necessary and sufficient conditions for the solvability of the H_∞ control problem have been obtained. In the continuous case, both state feedback and dynamic output feedback controllers have been designed. When parameter uncertainties appear in a system model, the notion of generalized quadratic stabilizability with an H_∞-norm bound has been proposed. It has been shown that generalized quadratic stabilizability with an H_∞-norm bound implies robust stabilizability with an H_∞-norm bound. Necessary and sufficient conditions for generalized quadratic stabilizability with an H_∞-norm bound have been obtained, and desired state feedback controllers have been constructed for both continuous and discrete singular systems.

6

Guaranteed Cost Control

6.1 Introduction

In the robust control of uncertain systems, it is usually desirable to design a controller which not only robustly stabilizes the uncertain system but also ensures an adequate level of performance. To this end, a design approach called guaranteed cost control is presented, in which an upper bound on the closed-loop value of a quadratic cost function can be guaranteed by using a fixed Lyapunov function. A great number of results on this topic have been presented in the literature [45, 46, 143, 174, 185, 196]. However, all these results were obtained in the context of state-space systems.

In this chapter, we consider the guaranteed cost control problem for linear singular systems with time-invariant norm-bounded parameter uncertainties. The purpose is the design of state feedback controllers such that the closed-loop system is admissible and a specified quadratic cost function has an upper bound for all admissible uncertainties. Sufficient conditions for the solvability of this problem are obtained. In both continuous and discrete cases, an LMI approach is developed to design desired state feedback controllers. Furthermore, examples are given to demonstrate the applicability of the proposed methods.

6.2 Continuous Systems

In this section, we will study the guaranteed cost control problem for uncertain continuous singular systems by using the LMI approach. The class of singular systems to be considered is described by

$$E\dot{x}(t) = (A + \Delta A)\,x(t) + (B + \Delta B)u(t), \tag{6.1}$$

where $x(t) \in \mathbb{R}^n$ is the state; $u(t) \in \mathbb{R}^m$ is the control input. The matrix $E \in \mathbb{R}^{n \times n}$ may be singular; we shall assume that $\mathrm{rank}(E) = r \leq n$. A and B are known real constant matrices with appropriate dimensions. ΔA and ΔB are unknown time-invariant matrices representing norm-bounded parameter uncertainties, which are assumed to be of the form

$$\begin{bmatrix} \Delta A\ \Delta B \end{bmatrix} = MF \begin{bmatrix} N_1\ N_2 \end{bmatrix}, \tag{6.2}$$

where M, N_1 and N_2 are known real constant matrices with appropriate dimensions. The uncertain matrix F satisfies

$$F^T F \leq I. \tag{6.3}$$

The initial condition of the singular system (6.1) is assumed to be

$$x(0) = \varphi(0).$$

Associated with the singular system (6.1) is the following cost function:

$$J = \int_0^\infty \left[x(t)^T Q_1 x(t) + u(t)^T Q_2 u(t) \right] dt, \tag{6.4}$$

where $Q_1 > 0$ and $Q_2 > 0$ are given constant matrices.

Now consider the following linear state feedback controller

$$u(t) = Kx(t), \quad K \in \mathbb{R}^{m \times n}. \tag{6.5}$$

Applying the controller in (6.5) to (6.1) results in the following closed-loop system:

$$E\dot{x}(t) = (A_c + \Delta A_c)\,x(t), \tag{6.6}$$

where

$$A_c = A + BK, \quad \Delta A_c = \Delta A + \Delta BK. \tag{6.7}$$

Then, the guaranteed cost control problem to be addressed in this section can be formulated as follows: given two constant matrices $Q_1 > 0$ and $Q_2 > 0$, design a state feedback controller in (6.5) such that the closed-loop system (6.6) is admissible and the cost function in (6.4) has an upper bound for all uncertainties satisfying (6.2) and (6.3). In this case, (6.5) is said to be a guaranteed cost state feedback controller.

Remark 6.1. It can be seen that when the closed-loop system (6.6) is admissible, the cost function (6.4) is well defined.   ◁

A sufficient condition for the solvability of the guaranteed cost control problem is given in the following theorem.

Theorem 6.1. *Consider uncertain singular system (6.1) and cost function (6.4). The guaranteed cost control problem is solvable if there exist matrices P, Y and a scalar $\epsilon > 0$ such that the following LMIs hold:*

$$EP = P^T E^T \geq 0, \tag{6.8}$$

$$\begin{bmatrix} \Psi + \epsilon M M^T & P^T N_1^T + Y^T N_2^T & P^T & Y^T \\ N_1 P + N_2 Y & -\epsilon I & 0 & 0 \\ P & 0 & -Q_1^{-1} & 0 \\ Y & 0 & 0 & -Q_2^{-1} \end{bmatrix} < 0, \tag{6.9}$$

where

$$\Psi = AP + P^T A^T + BY + Y^T B^T.$$

In this case, we can assume that the matrix P is nonsingular (if this is not the case, then we can choose some $\theta \in (0, 1)$ such that $\hat{P} = P + \theta \tilde{P}$ is nonsingular and satisfies (6.8) and (6.9), in which $\tilde{P}$ is any nonsingular matrix satisfying $E\tilde{P} = \tilde{P}^T E^T \geq 0$), and a desired guaranteed cost state feedback controller can be chosen as

$$u(t) = Y P^{-1} x(t), \tag{6.10}$$

and the corresponding cost function in (6.4) satisfies

$$J \leq \varphi(0)^T E^T P^{-1} \varphi(0). \tag{6.11}$$

Proof. Under the condition of the theorem, we first show the admissibility of the closed-loop system in (6.6). To this end, we apply Schur complement to (6.9) and obtain

$$\Psi + \epsilon M M^T + \epsilon^{-1} (N_1 P + N_2 Y)^T (N_1 P + N_2 Y)$$
$$+ P^T Q_1 P + Y^T Q_2 Y < 0. \tag{6.12}$$

As shown in the theorem, without loss of generality, we can assume that the matrix P is non-singular. Considering this, we denote

$$\bar{P} = P^{-1}.$$

Then, pre- and post-multiplying (6.12) by $\bar{P}^T$ and $\bar{P}$, respectively, provide

$$(A + BK)^T \bar{P} + \bar{P}^T (A + BK) + \epsilon \bar{P}^T M M^T \bar{P}$$
$$+ \epsilon^{-1} (N_1 + N_2 K)^T (N_1 + N_2 K) + Q_1 + K^T Q_2 K < 0, \tag{6.13}$$

where

$$K = Y \bar{P}.$$

Now, by Lemma 2.7, we have

$$\Delta A_c^T \bar{P} + \bar{P}^T \Delta A_c$$
$$= (N_1 + N_2 K)^T F^T M^T \bar{P} + \bar{P}^T M F (N_1 + N_2 K)$$
$$\leq \epsilon \bar{P}^T M M^T \bar{P} + \epsilon^{-1} (N_1 + N_2 K)^T (N_1 + N_2 K).$$

This together with (6.13) gives

$$(A_c + \Delta A_c)^T \bar{P} + \bar{P}^T (A_c + \Delta A_c) + Q_1 + K^T Q_2 K < 0, \tag{6.14}$$

which implies

$$(A_c + \Delta A_c)^T \bar{P} + \bar{P}^T (A_c + \Delta A_c) < 0. \tag{6.15}$$

Now, pre- and post-multiplying (6.8) by $\bar{P}^T$ and $\bar{P}$, respectively, we obtain

$$\bar{P}^T E = E^T \bar{P} \geq 0. \tag{6.16}$$

Noting (6.15) and (6.16), and then using Theorem 2.1, we have that the closed-loop system in (6.6) is admissible. Next, we will show that with controller (6.10), the cost function in (6.4) has an upper bound in (6.11). To this end, we set

$$V(x(t)) = x(t)^T E^T \bar{P} x(t).$$

Then, differentiating $V(x(t))$ with respect to time t along the solution of (6.6), we obtain

$$\dot{V}(x(t)) = x(t)^T \left[(A_c + \Delta A_c)^T \bar{P} + \bar{P}^T (A_c + \Delta A_c) \right] x(t)$$

By (6.14), it is easy to see that

$$\dot{V}(x(t)) < -x(t)^T (Q_1 + K^T Q_2 K) x(t) \tag{6.17}$$

Integrating both sides of (6.17) from 0 to any $T > 0$ gives

$$\int_0^T x(t)^T (Q_1 + K^T Q_2 K) x(t) dt \leq \varphi(0)^T E^T \bar{P} \varphi(0).$$

Therefore, the upper bound in (6.11) is satisfied. This completes the proof.$\square$

Theorem 6.1 provides a sufficient condition for the solvability of the guaranteed cost control problem. It is noted that the LMI in (6.8) is a non-strict one. Along a similar line as in the proof of Theorem 2.2, we can obtain the following solvability condition in terms of a strict LMI.

Theorem 6.2. *Consider uncertain singular system (6.1) and cost function (6.4). The guaranteed cost control problem is solvable if there exist matrices $P > 0$, Q, Y and a scalar $\epsilon > 0$ such that the following LMI holds:*

$$\begin{bmatrix} \hat{\Psi}(P,Q) + \epsilon MM^T & \Omega(P,Q)^T N_1^T + Y^T N_2^T & \Omega(P,Q)^T & Y^T \\ N_1\Omega(P,Q) + N_2 Y & -\epsilon I & 0 & 0 \\ \Omega(P,Q) & 0 & -Q_1^{-1} & 0 \\ Y & 0 & 0 & -Q_2^{-1} \end{bmatrix} < 0, \quad (6.18)$$

where

$$\hat{\Psi}(P,Q) = \Omega(P,Q)^T A^T + A\Omega(P,Q) + BY + Y^T B^T,$$
$$\Omega(P,Q) = PE^T + SQ,$$

and $S \in \mathbb{R}^{n \times (n-r)}$ is any matrix with full column rank and satisfies $ES = 0$. In this case, we can assume that the matrix $\Omega(P,Q)$ is nonsingular (if this is not the case, then we can choose some $\theta \in (0,1)$ such that $\hat{\Omega}(P,Q) = \Omega(P,Q) + \theta\tilde{P}$ is nonsingular and satisfies (6.18), in which $\tilde{P}$ is any nonsingular matrix satisfying $E\tilde{P} = \tilde{P}^T E^T \geq 0$), and a desired guaranteed cost state feedback controller can be chosen as

$$u(t) = Y\Omega(P,Q)^{-1} x(t), \quad (6.19)$$

and the corresponding cost function in (6.4) satisfies

$$J \leq \varphi(0)^T E^T \Omega(P,Q)^{-1} \varphi(0). \quad (6.20)$$

We now provide a numerical example to show the effectiveness of the proposed design method in this section.

Example 6.1. Consider an uncertain continuous singular system in (6.1) with

$$E = \begin{bmatrix} 10.5 & -3.5 & -3.5 \\ 27.6 & 19.3 & -18.7 \\ -1.2 & -8.6 & 3.4 \end{bmatrix}, \quad A = \begin{bmatrix} 7.45 & 8.05 & -5.05 \\ -2.81 & 22.11 & -7.71 \\ 6.22 & -1.22 & -1.18 \end{bmatrix},$$

$$B = \begin{bmatrix} 6 & 0.25 \\ 2 & 5.05 \\ 3.8 & -2.1 \end{bmatrix}, \quad M = \begin{bmatrix} 0.1 \\ -0.18 \\ 0.06 \end{bmatrix},$$

$$N_1 = \begin{bmatrix} 1.1 & 0 & -0.6 \end{bmatrix}, \quad N_2 = \begin{bmatrix} 0.1 & 0.27 \end{bmatrix}.$$

It can be verified that the nominal singular system is not admissible. In this example, the cost function is given in (6.4) with

$$Q_1 = \begin{bmatrix} 1 & 0.5 & 0.1 \\ 0.5 & 1 & 0.1 \\ 0.1 & 0.1 & 0.6 \end{bmatrix}, \quad Q_2 = \begin{bmatrix} 1.5 & 0.5 \\ 0.5 & 0.6 \end{bmatrix}.$$

Suppose the initial condition is

$$x(0) = \begin{bmatrix} 1.2 & 0.5 & -0.3 \end{bmatrix}.$$

In order to use Theorem 6.2 to solve the guaranteed cost control problem, we note that $\text{rank}(E) = 2$, thus we can choose

$$S = \begin{bmatrix} 0.4 \\ 0.3 \\ 0.9 \end{bmatrix},$$

which is with full column rank and satisfies $ES = 0$. Now, we obtain a set of solutions to the LMIs in (6.18) as follows:

$$P = \begin{bmatrix} 9.5372 & 7.0276 & 21.1903 \\ 7.0276 & 5.3503 & 15.9219 \\ 21.1903 & 15.9219 & 47.7609 \end{bmatrix},$$

$$Y = \begin{bmatrix} -3.2405 & -0.8789 & -2.1979 \\ 0.9011 & -2.0088 & 1.8932 \end{bmatrix},$$

$$Q = \begin{bmatrix} -0.6127 & -0.4019 & -0.1497 \end{bmatrix},$$

$$\epsilon = 10.4069.$$

Therefore, by Theorem 6.3, a guaranteed cost state feedback controller can be chosen as

$$u(t) = \begin{bmatrix} -0.3203 & 7.0407 & -3.2747 \\ 0.7695 & -6.9033 & 6.1725 \end{bmatrix} x(t).$$

By (6.20), it can be calculated that the corresponding cost function satisfies

$$J \leq 41.3995.$$

$\diamond$

6.3 Discrete Systems

Consider an uncertain linear discrete-time singular system with time-invariant norm-bounded parameter uncertainties, which is described by

$$Ex(k+1) = (A + \Delta A)x(k) + (B + \Delta B)u(k), \tag{6.21}$$

where $x(k) \in \mathbb{R}^n$ is the state; $u(k) \in \mathbb{R}^m$ is the control input. The matrix $E \in \mathbb{R}^{n \times n}$ may be singular; we shall assume that $\text{rank}(E) = r \leq n$. A and B are known real constant matrices with appropriate dimensions. ΔA and ΔB are time-invariant matrices representing norm-bounded parameter uncertainties, which are assumed to be of the form in (6.2) and (6.3).

The initial condition of the singular system (6.21) is assumed to be

$$x(0) = \varphi(0).$$

Associated with the singular system (6.21) is the following cost function:

$$J = \sum_{k=0}^{\infty} \left[x(k)^T Q_1 x(k) + u(k)^T Q_2 u(k) \right], \tag{6.22}$$

where $Q_1 > 0$ and $Q_2 > 0$ are given constant matrices.

Now consider the following linear state feedback controller

$$u(k) = Kx(k), \quad K \in \mathbb{R}^{m \times n}. \tag{6.23}$$

Applying the controller (6.23) to (6.21), we obtain the closed-loop system as follows:

$$Ex(k+1) = (A_c + \Delta A_c) x(k), \tag{6.24}$$

where

$$A_c = A + BK, \quad \Delta A_c = \Delta A + \Delta BK. \tag{6.25}$$

Similar to the continuous case, the guaranteed cost control problem to be addressed in this section can be formulated as follows: given two constant matrices $Q_1 > 0$ and $Q_2 > 0$, design a state feedback controller in (6.23) such that the closed-loop system in (6.24) is admissible and the cost function in (6.22) has an upper bound for all uncertainties satisfying (6.2) and (6.3). In this case, (6.25) is said to be a guaranteed cost state feedback controller.

We provide a sufficient condition for the solvability of the guaranteed cost control problem for discrete singular systems in the following theorem.

Theorem 6.3. *Consider uncertain discrete singular system (6.21) and cost function (6.22). The guaranteed cost control problem is solvable if there exist matrices P, G, Y and a scalar $\epsilon > 0$ satisfying*

$$E^T PE \geq 0, \tag{6.26}$$

and

$$\begin{bmatrix} \Upsilon + \epsilon MM^T & H_1^T + \epsilon MM^T & 0 \\ H_1 + \epsilon MM^T & \epsilon MM^T - 2G - 2G^T & G^T + I \\ 0 & G + I & P - 2I \\ G + I & 0 & 0 \\ H_2 & 0 & 0 \\ G & 0 & 0 \\ Y & 0 & 0 \end{bmatrix}$$

$$\begin{bmatrix} G^T + I & H_2^T & G^T & Y^T \\ 0 & 0 & 0 & 0 \\ 0 & 0 & 0 & 0 \\ -E^T P E - 2I & 0 & 0 & 0 \\ 0 & -\epsilon I & 0 & 0 \\ 0 & 0 & -Q_1^{-1} & 0 \\ 0 & 0 & 0 & -Q_2^{-1} \end{bmatrix} < 0, \tag{6.27}$$

where

$$\Upsilon = AG + G^T A^T + BY + Y^T B^T - G - G^T,$$
$$H_1 = AG + BY - G^T,$$
$$H_2 = N_1 G + N_2 Y.$$

In this case, a desired guaranteed cost state feedback controller can be chosen as

$$u(k) = Y G^{-1} x(k), \tag{6.28}$$

and the corresponding cost function in (6.22) satisfies

$$J \le \varphi(0)^T E^T P E \varphi(0). \tag{6.29}$$

Proof. First, from (6.27), it can be seen that

$$-G - G^T < 0.$$

Thus, the matrix G is non-singular and the state feedback controller in (6.28) is well defined. Now, applying this controller to (6.21) results in the closed-loop system in (6.24), and the matrices A_c and ΔA_c are given in (6.25) with

$$K = Y G^{-1}.$$

By Schur complement, it follows from (6.27) that

$$\begin{bmatrix} \Upsilon + G^T Q_1 G + Y^T Q_2 Y & H_1^T & 0 & G^T + I \\ H_1 & -2G - 2G^T & G^T + I & 0 \\ 0 & G + I & P - 2I & 0 \\ G + I & 0 & 0 & -E^T P E - 2I \end{bmatrix}$$
$$+ \epsilon^{-1} \begin{bmatrix} H_2^T \\ 0 \\ 0 \\ 0 \end{bmatrix} \begin{bmatrix} H_2^T \\ 0 \\ 0 \\ 0 \end{bmatrix}^T + \epsilon \begin{bmatrix} M \\ M \\ 0 \\ 0 \end{bmatrix} \begin{bmatrix} M \\ M \\ 0 \\ 0 \end{bmatrix}^T < 0. \tag{6.30}$$

Note that

$$\Delta A_c = M F (N_1 + N_2 K).$$

Then, by Lemma 2.7, we have

$$\begin{bmatrix} \Delta A_c G + G^T \Delta A_c^T & G^T \Delta A_c^T & 0 & 0 \\ \Delta A_c G & 0 & 0 & 0 \\ 0 & 0 & 0 & 0 \\ 0 & 0 & 0 & 0 \end{bmatrix}$$

$$= \begin{bmatrix} M \\ M \\ 0 \\ 0 \end{bmatrix} F \begin{bmatrix} H_2 & 0 & 0 & 0 \end{bmatrix}$$

$$+ \begin{bmatrix} H_2^T \\ 0 \\ 0 \\ 0 \end{bmatrix} F^T \begin{bmatrix} M^T & M^T & 0 & 0 \end{bmatrix}$$

$$\leq \epsilon^{-1} \begin{bmatrix} H_2^T \\ 0 \\ 0 \\ 0 \end{bmatrix} \begin{bmatrix} H_2^T \\ 0 \\ 0 \\ 0 \end{bmatrix}^T + \epsilon \begin{bmatrix} M \\ M \\ 0 \\ 0 \end{bmatrix} \begin{bmatrix} M \\ M \\ 0 \\ 0 \end{bmatrix}^T.$$

This together with (6.30) implies

$$\begin{bmatrix} \tilde{\Upsilon} + G^T Q_1 G + Y^T Q_2 Y & G^T (A_c + \Delta A_c)^T - G \\ (A_c + \Delta A_c) G - G^T & -2G - 2G^T \\ 0 & G + I \\ G + I & 0 \end{bmatrix}$$
$$\begin{matrix} 0 & G^T + I \\ G^T + I & 0 \\ P - 2I & 0 \\ 0 & -E^T P E - 2I \end{matrix} \Bigg] < 0, \qquad (6.31)$$

where

$$\tilde{\Upsilon} = (A_c + \Delta A_c) G + G^T (A_c + \Delta A_c)^T - G - G^T. \qquad (6.32)$$

Pre- and post-multiplying (6.31) by

$$\begin{bmatrix} I & 0 & 0 & G^T \\ 0 & I & 0 & 0 \\ 0 & 0 & I & 0 \end{bmatrix},$$

and its transpose, respectively, we obtain

$$\begin{bmatrix} \hat{\Upsilon} + G^T Q_1 G + Y^T Q_2 Y & G^T (A_c + \Delta A_c)^T - G & 0 \\ (A_c + \Delta A_c) G - G^T & -2G - 2G^T & G^T + I \\ 0 & G + I & P - 2I \end{bmatrix} < 0, \qquad (6.33)$$

where

$$\hat{\Upsilon} = (A_c + \Delta A_c) G + G^T (A_c + \Delta A_c)^T - G^T E^T P E G.$$

Pre- and post-multiplying (6.33) by

$$\begin{bmatrix} I & 0 & 0 \\ 0 & I & G^T \end{bmatrix},$$

and its transpose, respectively, result in

$$\begin{bmatrix} \hat{\Upsilon} + G^T Q_1 G + Y^T Q_2 Y & G^T (A_c + \Delta A_c)^T - G \\ (A_c + \Delta A_c) G - G^T & G^T PG - G - G^T \end{bmatrix} < 0.$$

Now, pre- and post-multiplying this inequality by

$$\begin{bmatrix} G^{-T} (A_c + \Delta A_c)^T G^{-T} \end{bmatrix}$$

and its transpose, respectively, we have

$$(A_c + \Delta A_c)^T P (A_c + \Delta A_c) - E^T PE + Q_1 + K^T Q_2 K < 0, \qquad (6.34)$$

which implies

$$(A_c + \Delta A_c)^T P (A_c + \Delta A_c) - E^T PE < 0.$$

Noting this and (6.26), and then using Theorem 2.3, we have that the closed-loop system (6.24) is admissible. To show the upper bound in (6.29), we define

$$V(x(k)) = x(k)^T E^T PEx(k).$$

Then, it can be calculated that

$$V(x(k+1)) - V(x(k))$$
$$= x(k)^T \left[(A_c + \Delta A_c)^T P (A_c + \Delta A_c) - E^T PE \right] x(k). \qquad (6.35)$$

From this and (6.34), it is easy to have

$$V(x(k+1)) - V(x(k)) \leq -x(k)^T (Q_1 + K^T Q_2 K) x(k). \qquad (6.36)$$

Summing up both sides of (6.36) from 0 to any $T > 0$ gives

$$\sum_{k=0}^{T} x(k)^T (Q_1 + K^T Q_2 K) x(k) \leq \varphi(0)^T E^T PE\varphi(0).$$

Therefore, the upper bound in (6.29) is satisfied. This completes the proof.□

Now, we provide an example to show the applicability of Theorem 6.3.

Example 6.2. Consider an uncertain discrete singular system in (6.21) with

$$E = \begin{bmatrix} 3 & 0 & 0 \\ 0 & 5 & 0 \\ 0 & 0 & 0 \end{bmatrix}, \quad A = \begin{bmatrix} 0.2 & 0.1 & 0.3 \\ -0.2 & 0.1 & 0.2 \\ 0 & 0 & 0 \end{bmatrix},$$

$$B = \begin{bmatrix} 1 & 0 & -0.8 \\ 0 & -0.6 & 0.5 \\ 0.2 & 1 & -1 \end{bmatrix}, \quad M = \begin{bmatrix} 0.1 \\ 0 \\ -0.1 \end{bmatrix},$$

$$N_1 = \begin{bmatrix} 0.2 & 0.1 & 0.1 \end{bmatrix}, \quad N_2 = \begin{bmatrix} -0.1 & 0 & 0.1 \end{bmatrix}.$$

The cost function is given in (6.22) with

$$Q_1 = \begin{bmatrix} 0.2 & 0.1 & 0.1 \\ 0.1 & 0.2 & 0.1 \\ 0.1 & 0.1 & 0.2 \end{bmatrix}, \quad Q_2 = \begin{bmatrix} 0.2 & 0.2 & 0.1 \\ 0.2 & 0.3 & 0.1 \\ 0.1 & 0.1 & 0.2 \end{bmatrix}.$$

In this example, we suppose the initial condition is

$$x(0) = \begin{bmatrix} 0.5 & -1.2 & 0.1 \end{bmatrix}.$$

It can be verified that in this example the nominal singular system is not admissible. To solve the guaranteed cost control problem, we solve the LMIs in (6.26) and (6.27) and obtain a set of solutions as follows:

$$P = \begin{bmatrix} 0.4520 & -0.0083 & -0.0231 \\ -0.0083 & 0.4382 & 0.3867 \\ -0.0231 & 0.3867 & -113.2149 \end{bmatrix},$$

$$G = \begin{bmatrix} 1.3494 & -0.1839 & -0.0533 \\ -0.1208 & 1.7579 & -0.3104 \\ -0.0596 & -0.3491 & 1.0301 \end{bmatrix},$$

$$Y = \begin{bmatrix} -0.3093 & -0.2093 & 0.5054 \\ 0.1532 & 0.2709 & -1.0077 \\ 0.1522 & -0.1400 & 0.8539 \end{bmatrix},$$

$$\epsilon = 40.3879.$$

Therefore, by Theorem 6.3, a guaranteed cost state feedback controller can be chosen as

$$u(k) = \begin{bmatrix} -0.2131 & -0.0491 & 0.4648 \\ 0.0669 & -0.0345 & -0.9852 \\ 0.1610 & 0.1100 & 0.8704 \end{bmatrix} x(k).$$

By (6.29), it can be shown that the corresponding cost function satisfies

$$J \leq 16.9414.$$

◇

6.4 Conclusion

This chapter has studied the problem of guaranteed cost control for both continuous and discrete singular systems with time-invariant norm-bounded parameter uncertainties. State feedback controllers have been designed such that the closed-loop system is admissible and a quadratic cost function has an upper bound for all admissible uncertainties. In both the continuous and discrete cases, an LMI approach has been developed to design desired state feedback controllers. Examples have been provided to demonstrate the applicability of the proposed methods.

7

Positive Real Control

7.1 Introduction

The notion of positive realness has played an important role in control and system theory [5, 70, 167]. A well-known fact in robust and nonlinear control is that the positive realness of a certain loop transfer function will guarantee the overall stability of a feedback system if uncertainty or nonlinearity can be characterized by a positive real system [167]. This has motivated the study of the positive real control problem. In the context of state-space systems, results on positive real control can be found in [153, 175], and the references therein.

In this chapter, we consider the positive real control problem for singular systems in both the continuous and discrete cases. The aim is to design a state feedback controller such that the closed-loop system is regular, impulse-free (for continuous singular systems) or causality (for discrete singular systems), and extended strictly positive real. First, versions of the positive real lemma are proposed in terms of LMIs, which can be viewed as extensions of positive real lemmas for state-space systems to singular systems. Based on the proposed positive real lemmas, necessary and sufficient conditions for the solvability of the positive real control problem are obtained. For continuous singular systems, a desired state feedback controller can be constructed by solving an LMI, while for discrete singular systems, a desired state feedback controller can be constructed by solving a matrix inequality.

7.2 Continuous Systems

In this section, we consider the positive real control problem for continuous singular systems. The class of continuous singular systems to be considered are described by

$$E\dot{x}(t) = Ax(t) + B_1 u(t) + Bw(t), \tag{7.1}$$

$$z(t) = Cx(t) + B_2 u(t) + Dw(t), \tag{7.2}$$

where $x(t) \in \mathbb{R}^n$ is the state; $u(t) \in \mathbb{R}^m$ is the control input; $w(t) \in \mathbb{R}^l$ is the exogenous input; $z(t) \in \mathbb{R}^s$ is the controlled output. In this singular system, we assume $l = s$. The matrix $E \in \mathbb{R}^{n \times n}$ may be singular; we shall assume that rank$(E) = r \leq n$. A, B, B_1, B_2, C and D are known real constant matrices with appropriate dimensions.

Considering that the positive real lemma has played an important role in solving the positive real control problem for state-space systems, we first provide a version of the positive real lemma for continuous singular systems before studying the positive real control problem. To this end, we consider the singular system in (7.1) and (7.2) with $u(t) = 0$; that is,

$$E\dot{x}(t) = Ax(t) + Bw(t), \tag{7.3}$$

$$z(t) = Cx(t) + Dw(t). \tag{7.4}$$

Then the transfer function of the above system is as follows:

$$G(s) = C\left(sE - A\right)^{-1} B + D. \tag{7.5}$$

For the continuous singular system in (7.3) and (7.4), we introduce the following definition of positive realness.

Definition 7.1. [153, 203]

(1) The singular system in (7.3) and (7.4) is said to be positive real (PR) if its transfer function $G(s)$ is analytic in $Re(s) > 0$ and satisfies $G(s) + G(s)^* \geq 0$ for $Re(s) > 0$.

(2) The singular system in (7.3) and (7.4) is said to be strictly positive real (SPR) if its transfer function $G(s)$ is analytic in $Re(s) \geq 0$ and satisfies $G(j\theta) + G(-j\theta)^T > 0$ for $\theta \in [0, \infty)$.

(3) The singular system in (7.3) and (7.4) is said to be extended strictly positive real (ESPR) if it is SPR and $G(j\infty) + G(-j\infty)^T > 0$.

Now, we present a version of the positive real lemma for continuous singular systems in the following theorem.

Theorem 7.1. *Suppose $D + D^T > 0$. The continuous singular system in (7.3) and (7.4) is admissible and ESPR if and only if there exists a matrix P such that the following LMIs hold:*

$$E^T P = P^T E \geq 0, \tag{7.6}$$

$$\begin{bmatrix} A^T P + P^T A & C^T - P^T B \\ C - B^T P & -(D + D^T) \end{bmatrix} < 0. \tag{7.7}$$

To prove this theorem, we need the following lemmas.

Lemma 7.1. [153] *The continuous state-space system described by*

$$\dot{x}(t) = Ax(t) + Bw(t),$$
$$z(t) = Cx(t) + Dw(t),$$

is stable and ESPR if and only if there exists a matrix $P > 0$ such that the following LMI holds:

$$\begin{bmatrix} A^T P + PA & C^T - PB \\ C - B^T P & -(D + D^T) \end{bmatrix} < 0.$$

Lemma 7.2. [55, 84] *Given a symmetric matrix $\Xi \in \mathbb{R}^{n \times n}$ and two matrices $\Gamma \in \mathbb{R}^{n \times m}$ and $\Pi \in \mathbb{R}^{k \times n}$ with $\mathrm{rank}(\Gamma) < n$ and $\mathrm{rank}(\Pi) < n$. Consider the problem of finding some matrix Θ such that*

$$\Xi + \Gamma \Theta \Pi + (\Gamma \Theta \Pi)^T < 0. \tag{7.8}$$

Then (7.8) is solvable for Θ if and only if

$$\Gamma^\perp \Xi \Gamma^{\perp T} < 0, \quad \left(\Pi^T\right)^\perp \Xi \left(\Pi^T\right)^{\perp T} < 0.$$

Proof of Theorem 7.1. (*Sufficiency*) Suppose that there exists a matrix P such that the LMIs in (7.6) and (7.7) hold. Then, it is easy to see that

$$A^T P + P^T A < 0.$$

Noting this, (7.6) and then applying Theorem 2.1, we have that the continuous singular system in (7.3) is admissible. Next, we shall show that under the condition of the theorem, the singular system in (7.3) and (7.4) is ESPR. To this end, we apply Schur complement to (7.7) and obtain

$$A^T P + P^T A + \left(C^T - P^T B\right)\left(D + D^T\right)^{-1}\left(C - B^T P\right) < 0,$$

which implies that there exists a constant matrix $Z > 0$ such that

$$A^T P + P^T A + \left(C^T - P^T B\right)\left(D + D^T\right)^{-1}\left(C - B^T P\right) + Z < 0.$$

With this and (7.6), it can be verified that

$$-\left(-j\theta E - A\right)^T P - P^T \left(j\theta E - A\right) + W < 0, \tag{7.9}$$

where

$$W = \left(C^T - P^T B\right)\left(D + D^T\right)^{-1}\left(C - B^T P\right) + Z.$$

Since the singular system in (7.3) is admissible, we have that $j\theta E - A$ is invertible for all $\theta \in \mathbb{R}$. Then, pre- and post-multiplying (7.9) by $\left(-j\theta E - A\right)^{-T}$ and $\left(j\theta E - A\right)^{-1}$, respectively, result in

$$-P\left(j\theta E - A\right)^{-1} - \left(-j\theta E - A\right)^{-T} P^T$$
$$+\left(-j\theta E - A\right)^{-T} W \left(j\theta E - A\right)^{-1} < 0.$$

That is,

$$P\left(j\theta E - A\right)^{-1} + \left(-j\theta E - A\right)^{-T} P^T > \left(-j\theta E - A\right)^{-T} W \left(j\theta E - A\right)^{-1}.$$

Using this inequality, we can deduce that

$$\begin{aligned}
&G(j\theta) + G(-j\theta)^T \\
&= C\left(j\theta E - A\right)^{-1} B + B^T \left(-j\theta E - A\right)^{-T} C^T + \left(D + D^T\right) \\
&= \left(C - B^T P\right)\left(j\theta E - A\right)^{-1} B + B^T \left(-j\theta E - A\right)^{-T} \left(C^T - P^T B\right) \\
&\quad + \left(D + D^T\right) + B^T \left[P\left(j\theta E - A\right)^{-1} + \left(-j\theta E - A\right)^{-T} P^T\right] B \\
&\geq \left(C - B^T P\right)\left(j\theta E - A\right)^{-1} B + B^T \left(-j\theta E - A\right)^{-T} \left(C^T - P^T B\right) \\
&\quad + \left(D + D^T\right) + B^T \left(-j\theta E - A\right)^{-T} W \left(j\theta E - A\right)^{-1} B \\
&= \left[B^T \left(-j\theta E - A\right)^{-T} + \left(C - B^T P\right) W^{-1}\right] \\
&\quad \times W \left[\left(j\theta E - A\right)^{-1} B + W^{-1}\left(C^T - P^T B\right)\right] \\
&\quad + \left(D + D^T\right) - \left(B^T P - C\right) W^{-1}\left(P^T B - C^T\right) \\
&\geq D + D^T - \left(B^T P - C\right) W^{-1}\left(P^T B - C^T\right). \tag{7.10}
\end{aligned}$$

Note that

$$W - \left(P^T B - C^T\right)\left(D + D^T\right)^{-1}\left(B^T P - C\right) = Z > 0,$$

which, by Schur complement, implies

$$\begin{bmatrix} W & P^T B - C^T \\ B^T P - C & D + D^T \end{bmatrix} > 0.$$

Applying Schur complement to this inequality, we have

$$D + D^T - \left(B^T P - C \right) W^{-1} \left(P^T B - C^T \right) > 0.$$

This together with (7.10) implies that for all $\theta \in \mathbb{R}$,

$$G(j\theta) + G(-j\theta)^T > 0,$$

and

$$G(j\infty) + G(-j\infty)^T > 0.$$

Therefore, by Definition 7.1, we have that the continuous singular system in (7.3) and (7.4) is ESPR.

(*Necessity*) Suppose that the continuous singular system in (7.3) and (7.4) is admissible and ESPR. Then, there exist nonsingular matrices M and N such that

$$E = M \begin{bmatrix} I & 0 \\ 0 & 0 \end{bmatrix} N, \quad A = M \begin{bmatrix} A_1 & 0 \\ 0 & I \end{bmatrix} N, \tag{7.11}$$

where $A_1 \in \mathbb{R}^{r \times r}$ is stable. Write

$$CN^{-1} = \begin{bmatrix} C_1 & C_2 \end{bmatrix}, \quad M^{-1}B = \begin{bmatrix} B_1 \\ B_2 \end{bmatrix},$$

where the partition is compatible with that of A in (7.11). Then, we have

$$G(s) = C_1 \left(sI - A_1 \right)^{-1} B_1 + D - C_2 B_2.$$

Thus, it is easy to see that the admissibility and ESPR of the singular system in (7.3) and (7.4) imply the state-space described by

$$\dot{\xi}(t) = A_1 \xi(t) + B_1 \tilde{w}(t),$$
$$\tilde{z}(t) = C_1 \xi(t) + (D - C_2 B_2) \tilde{w}(t),$$

is stable and ESPR, where $\xi(t) \in \mathbb{R}^r$. Therefore, by Lemma 7.1, it can be seen that there exists a matrix $P_1 > 0$ such that

$$\begin{bmatrix} A_1^T P_1 + P_1 A & C_1^T - P_1^T B_1 \\ C_1 - B_1^T P_1 & C_2 B_2 + B_2^T C_2^T - \left(D + D^T \right) \end{bmatrix} < 0. \tag{7.12}$$

Now, for a fixed $P_1 > 0$ satisfying (7.12), we claim that there exist matrices P_2 and P_3 such that

$$\Omega + \Psi \tilde{P} \Phi + (\Psi \tilde{P} \Phi)^T < 0, \tag{7.13}$$

where

$$\Omega = \begin{bmatrix} A_1^T P_1 + P_1 A & 0 & C_1^T - P_1^T B_1 \\ 0 & 0 & C_2^T \\ C_1 - B_1^T P_1 & C_2 - (D + D^T) \end{bmatrix}, \quad \Psi = \begin{bmatrix} 0 \\ I \\ -B_2^T \end{bmatrix},$$

$$\Phi = \begin{bmatrix} I & 0 & 0 \\ 0 & I & 0 \end{bmatrix}, \quad \tilde{P} = \begin{bmatrix} P_2 & P_3 \end{bmatrix}.$$

To show this, we note that

$$\Psi^\perp = \begin{bmatrix} I & 0 & 0 \\ 0 & B_2^T & I \end{bmatrix}, \quad (\Phi^T)^\perp = \begin{bmatrix} 0 & 0 & I \end{bmatrix}.$$

Then, it is easy to obtain that

$$\Psi^\perp \Omega \left(\Psi^\perp \right)^T = \begin{bmatrix} A_1^T P_1 + P_1 A & C_1^T - P_1^T B_1 \\ C_1 - B_1^T P_1 & C_2 B_2 + B_2^T C_2^T - (D + D^T) \end{bmatrix},$$

$$(\Phi^T)^\perp \Omega \left((\Phi^T)^\perp \right)^T = - (D + D^T).$$

Considering (7.12) and $D + D^T > 0$, we have

$$\Psi^\perp \Omega \left(\Psi^\perp \right)^T < 0, \quad (\Phi^T)^\perp \Omega \left((\Phi^T)^\perp \right)^T < 0.$$

Therefore, by Lemma 7.2, it can be seen that there exist matrices P_2 and P_3 such that (7.13) is feasible. Now, choose

$$P = M^{-T} \begin{bmatrix} P_1 & 0 \\ P_2 & P_3 \end{bmatrix} N. \tag{7.14}$$

Then, with the matrix given in (7.14), it can be verified that (7.13) can be rewritten as

$$\begin{bmatrix} N^{-T} \left(A^T P + P^T A \right) N^{-1} & N^{-T} \left(C^T - P^T B \right) \\ (C - B^T P) N^{-1} & - (D + D^T) \end{bmatrix} < 0.$$

Pre- and post-multiplying this inequality by $\mathrm{diag}(N^T, I)$ and its transpose, respectively, we have (7.7). Therefore, the matrix P given in (7.14) satisfies (7.6) and (7.7). This completes the proof. □

Remark 7.1. In the case when $E = I$, it is easy to see that the positive real lemma in Theorem 7.1 reduces to that in Lemma 7.1 for state-space systems. Therefore, Theorem 7.1 can be viewed as an extension of the positive real lemma for continuous state-space systems to continuous singular systems. ◁

Based on Theorem 7.1, we can also provide the following version of the positive real lemma for continuous singular systems.

Corollary 7.1. *Suppose $D+D^T > 0$. The continuous singular system in (7.3) and (7.4) is admissible and ESPR if and only if there exists a matrix P such that the following LMIs hold:*

$$EP = P^T E^T \geq 0,$$

$$\begin{bmatrix} AP + P^T A^T & B - P^T C^T \\ B^T - CP & -(D + D^T) \end{bmatrix} < 0.$$

Theorem 7.1 and Corollary 7.1 present two versions of the positive real lemma for continuous singular systems. It is noted that these positive real lemmas involve non-strict LMIs. Now, along a similar line to the derivation of Theorem 2.2, and by Theorem 7.1 and Corollary 7.1, we can obtain the following two versions of the bounded real lemma involving strict LMIs.

Theorem 7.2. *Suppose $D+D^T > 0$. The continuous singular system in (7.3) and (7.4) is admissible and ESPR if and only if there exist matrices $P > 0$ and Q such that the following LMI holds:*

$$\begin{bmatrix} A^T (PE + SQ) + (PE + SQ)^T A & C^T - (PE + SQ)^T B \\ C - B^T (PE + SQ) & -(D + D^T) \end{bmatrix} < 0. \qquad (7.15)$$

where $S \in \mathbb{R}^{n \times (n-r)}$ is any matrix with full column rank and satisfies $E^T S = 0$.

Corollary 7.2. *Suppose $D+D^T > 0$. The continuous singular system in (7.3) and (7.4) is admissible and ESPR if and only if there exist matrices $P > 0$ and Q such that the following LMI holds:*

$$\begin{bmatrix} A (PE^T + SQ) + (PE^T + SQ)^T A^T & B - (PE^T + SQ)^T C^T \\ B^T - C (PE^T + SQ) & -(D + D^T) \end{bmatrix} < 0.$$

where $S \in \mathbb{R}^{n \times (n-r)}$ is any matrix with full column rank and satisfies $ES = 0$.

Now, we investigate the positive real control problem for continuous singular systems. To this end, we consider the following state feedback controller

$$u(t) = Kx(t), \quad K \in \mathbb{R}^{m \times n}. \qquad (7.16)$$

Applying this controller to the singular system in (7.1) and (7.2) results in the closed-loop system as

$$E\dot{x}(t) = (A + B_1 K) x(t) + Bw(t), \qquad (7.17)$$

$$z(t) = (C + B_2 K) x(t) + Dw(t), \qquad (7.18)$$

Then, the positive real control problem to be addressed is the design of a state feedback controller (7.16) such that the closed-loop system in (7.17) and (7.18) is admissible and ESPR.

By resorting to Corollary 7.2 and following a similar line to the derivation of Theorem 3.1, we have the solution to the positive real control problem as follows.

Theorem 7.3. *Given a continuous singular system in (7.1) and (7.2) with*

$$D + D^T > 0.$$

There exists a state feedback controller in (7.16) such that the closed-loop system in (7.17) and (7.18) is admissible and ESPR if and only if there exist matrices $P > 0$, Q and Y such that

$$\begin{bmatrix} \Pi\,(P,Q) & B - \Omega\,(P,Q)^T\,C^T - Y^T B_2^T \\ B^T - C\Omega\,(P,Q) - B_2 Y & -\,(D + D^T) \end{bmatrix} < 0, \qquad (7.19)$$

where

$$\Pi\,(P,Q) = \Omega\,(P,Q)^T\,A^T + A\Omega\,(P,Q) + B_1 Y + Y^T B_1^T,$$
$$\Omega\,(P,Q) = PE^T + SQ,$$

and $S \in \mathbb{R}^{n \times (n-r)}$ is any matrix with full column rank and satisfies $ES = 0$. In this case, we can assume that the matrix $\Omega\,(P,Q)$ is nonsingular (if this is not the case, then we can choose some $\theta \in (0,1)$ such that $\hat{\Omega}\,(P,Q) = \Omega\,(P,Q) + \theta\tilde{P}$ is nonsingular and satisfies (7.19), in which $\tilde{P}$ is any nonsingular matrix satisfying $E\tilde{P} = \tilde{P}^T E^T \geq 0$), then a desired state feedback controller can be chosen as

$$u(t) = Y\Omega\,(P,Q)^{-1}\,x(t). \qquad (7.20)$$

The following example demonstrates the effectiveness of the proposed design method in this section.

Example 7.1. Consider a continuous singular system in (7.1) and (7.2) with parameters as follows:

$$E = \begin{bmatrix} 13.2 & 2.2 \\ -30 & -5 \end{bmatrix}, \quad A = \begin{bmatrix} 14.3 & 4.4 \\ -32.5 & 10 \end{bmatrix},$$

$$B_1 = \begin{bmatrix} 1 & -1.8 \\ -2 & 4 \end{bmatrix}, \quad B = \begin{bmatrix} -2 & 0.5 \\ 4.5 & -1.25 \end{bmatrix},$$

$$C = \begin{bmatrix} 8.2 & 3.2 \\ 5 & -1 \end{bmatrix}, \quad B_2 = \begin{bmatrix} 0 & 1 \\ -1 & 2 \end{bmatrix},$$

$$D = \begin{bmatrix} 1.2 & 0.5 \\ 0 & 1 \end{bmatrix}.$$

Then, it can be verified that the unforced system of this singular system is not admissible and ESPR. The purpose is to design a state feedback controller such that the closed-loop system is admissible and ESPR. For this purpose, we first note that $D + D^T > 0$. Since rank$(E) = 1$, we can choose

$$S = \begin{bmatrix} -0.1818 & 1.0909 \end{bmatrix}^T,$$

which is with full column rank and satisfies $ES = 0$. Now, we found that the LMI in (7.19) is feasible and a set of solutions is as follows:

$$P = \begin{bmatrix} 2.7059 & -16.2140 \\ -16.2140 & 97.2877 \end{bmatrix},$$

$$Y = \begin{bmatrix} -83.8847 & -284.4653 \\ -284.4653 & -93.4366 \end{bmatrix},$$

$$Q = \begin{bmatrix} 12.7887 & 49.4152 \end{bmatrix}.$$

Therefore, by Theorem 7.3, the positive real control problem is solvable and a desired state feedback controller can be chosen as in (7.20), which can be calculated as

$$u(t) = \begin{bmatrix} -131.3497 & -27.4410 \\ -45.5801 & -9.4245 \end{bmatrix} x(t).$$

◇

7.3 Discrete Systems

In this section, we consider the positive real control problem for discrete singular systems. The discrete singular system to be considered is described by

$$Ex(k + 1) = Ax(k) + B_1u(k) + Bw(k), \qquad (7.21)$$
$$z(k) = Cx(k) + Dw(k), \qquad (7.22)$$

where $x(k) \in \mathbb{R}^n$ is the state; $u(k) \in \mathbb{R}^m$ is the control input; $w(k) \in \mathbb{R}^l$ is the disturbance input; $z(k) \in \mathbb{R}^s$ is the controlled output. In this system, we assume $l = s$. The matrix $E \in \mathbb{R}^{n \times n}$ may be singular; we shall assume that rank$(E) = r \leq n$. A, B, B_1, C and D are known real constant matrices with appropriate dimensions.

Similar to the continuous case, we first provide a version of the positive real lemma for discrete singular system before investigating the positive real control problem. To this end, we consider the singular system in (7.21) and (7.22) with $u(k) = 0$; that is,

$$Ex(k+1) = Ax(k) + Bw(k), \qquad (7.23)$$
$$z(k) = Cx(k) + Dw(k). \qquad (7.24)$$

Then the transfer function of the above system is as follows:

$$G(z) = C\,(zE - A)^{-1}\,B + D. \qquad (7.25)$$

For the discrete singular system in (7.23) and (7.24), we introduce the following concept of positive realness.

Definition 7.2. [203]

(I) *The singular system in (7.23) and (7.24) is said to be positive real (PR) if its transfer function $G(z)$ is analytic in $|z| > 1$ and satisfies $G(z) + G^*(z) \geq 0$ for $|z| > 1$.*

(II) *The singular system in (7.23) and (7.24) is said to be strictly positive real (SPR) if its transfer function $G(z)$ is analytic in $|z| \geq 1$ and satisfies $G(e^{j\theta}) + G^*(e^{j\theta}) > 0$ for $\theta \in [0, 2\pi)$.*

(III) *The singular system in (7.23) and (7.24) is said to be extended strictly positive real (ESPR) if it is SPR and $G(\infty) + G(\infty)^T > 0$.*

We now provide a version of positive real lemma for discrete singular systems in the following theorem.

Theorem 7.4. *The discrete singular system in (7.23) and (7.24) is admissible and ESPR if and only if there exists a matrix $P = P^T$ such that the following LMIs hold:*

$$E^T P E \geq 0, \qquad (7.26)$$

$$\begin{bmatrix} A^T P A - E^T P E & C^T - A^T P B \\ C - B^T P A & -\left(D^T + D - B^T P B\right) \end{bmatrix} < 0. \qquad (7.27)$$

To prove this theorem, we need the following lemma.

Lemma 7.3. [71] *The discrete state-space system described by*

$$x(k+1) = Ax(k) + Bw(k),$$
$$z(k) = Cx(k) + Dw(k),$$

is stable and ESPR if and only if there exists a matrix $P > 0$ such that

$$\begin{bmatrix} A^T P A - P & C^T - A^T P B \\ C - B^T P A & -\left(D^T + D - B^T P B\right) \end{bmatrix} < 0.$$

Proof of Theorem 7.4. (*Sufficiency*) Suppose that there exists a matrix $P = P^T$ such that the LMIs in (7.26) and (7.27) hold. Then, it is easy to see

$$A^T P A - P < 0.$$

Noting this and (7.26), and then using Theorem 2.3, we have that the discrete singular system in (7.23) is admissible, which further implies that the transfer function $G(z)$ in (7.25) is analytic in $|z| \geq 1$. Next, we shall show that under the condition of the theorem, the singular system in (7.3) and (7.4) is ESPR. To this end, we first establish

$$G(\infty) + G(\infty)^T > 0. \tag{7.28}$$

Note that the admissibility of the singular system in (7.23) and (7.24) implies that there exist non-singular matrices M and N such that

$$MEN = \begin{bmatrix} I & 0 \\ 0 & 0 \end{bmatrix}, \quad MAN = \begin{bmatrix} A_1 & 0 \\ 0 & I \end{bmatrix}, \tag{7.29}$$

$$CN = \begin{bmatrix} C_1 & C_2 \end{bmatrix}, \quad MB = \begin{bmatrix} B_1 \\ B_2 \end{bmatrix}, \tag{7.30}$$

where $A_1 \in \mathbb{R}^{r \times r}$, $B_1 \in \mathbb{R}^{r \times s}$, $B_2 \in \mathbb{R}^{(n-r) \times s}$, $C_1 \in \mathbb{R}^{s \times r}$, $C_2 \in \mathbb{R}^{s \times (n-r)}$. Then, it is easy to see

$$G(z) = C_1 (zI - A_1)^{-1} B_1 + D - C_2 B_2. \tag{7.31}$$

Denote

$$M^{-T} P M^{-1} = \begin{bmatrix} P_1 & P_2 \\ P_2^T & P_3 \end{bmatrix}, \tag{7.32}$$

where the matrix partition is compatible with those in (7.29) and (7.30). Then, by (7.26), (7.29) and (7.30), we have

$$P_1 \geq 0. \tag{7.33}$$

Now, pre- and post-multiplying (7.27) by $\mathrm{diag}\left(N^T, I\right)$ and its transpose, respectively, we obtain

$$\begin{bmatrix} * & * & * \\ * & P_3 & C_2^T - P_2^T B_1 - P_3 B_2 \\ * & C_2 - B_1^T P_2 - B_2^T P_3 & U \end{bmatrix} < 0, \tag{7.34}$$

where

$$U = -\left(D^T + D\right) + B_1^T P_1 B_1 + B_2^T P_2^T B_1 + B_1^T P_2 B_2 + B_2^T P_3 B_2,$$

and $*$ represents a matrix which is not related to the following development. Now, pre- and post-multiplying (7.34) by

$$\begin{bmatrix} 0 & B_2^T & I \end{bmatrix},$$

and its transpose, respectively, result in

$$C_2 B_2 + B_2^T C_2^T - \left(D^T + D\right) + B_1^T P_1 B_1 < 0.$$

This together with (7.33) implies

$$C_2 B_2 + B_2^T C_2^T - \left(D^T + D\right) < 0.$$

With this and (7.31), it is easy to see that (7.28) holds. Next we shall show

$$G(e^{j\theta}) + G^*(e^{j\theta}) > 0, \tag{7.35}$$

for all $\theta \in [0, 2\pi)$. For this purpose, we denote

$$V = D + D^T - B^T P B.$$

Then, by Schur complement, it follows from (7.27) that

$$A^T P A - E^T P E + \left(C^T - A^T P B\right) V^{-1} \left(C - B^T P A\right) < 0.$$

This implies that there exists a matrix $Q_1 > 0$ such that

$$A^T P A - E^T P E + \left(C^T - A^T P B\right) V^{-1} \left(C - B^T P A\right) + Q_1 < 0. \tag{7.36}$$

Now, set

$$\Psi(j\theta) = e^{j\theta} E - A,$$
$$H = \left(C^T - A^T P B\right) V^{-1} \left(C - B^T P A\right) + Q_1.$$

Since the singular system in (7.23) is admissible, we have that $\Psi(j\theta)$ is invertible for all $\theta \in [0, 2\pi)$. Then, pre- and post-multiplying (7.36) by $B^T \Psi(-j\theta)^{-T}$ and $\Psi(j\theta)^{-1} B$, respectively, we have that for all $\theta \in [0, 2\pi)$,

$$B^T \Psi(-j\theta)^{-T} \left(A^T P A - E^T P E\right) \Psi(j\theta)^{-1} B$$
$$+ B^T \Psi(-j\theta)^{-T} H \Psi(j\theta)^{-1} B \leq 0. \tag{7.37}$$

Observe that for all $\theta \in [0, 2\pi)$,

$$B^T P B + B^T P A \Psi(j\theta)^{-1} B + B^T \Psi(-j\theta)^{-T} A^T P B$$
$$+ B^T \Psi(-j\theta)^{-T} \left(A^T P A - E^T P E\right) \Psi(j\theta)^{-1} B = 0.$$

Substituting this into (7.37) gives

$$B^T P B \geq -B^T P A \Psi(j\theta)^{-1} B - B^T \Psi(-j\theta)^{-T} A^T P B$$
$$+ B^T \Psi(-j\theta)^{-T} H \Psi(j\theta)^{-1} B,$$

for all $\theta \in [0, 2\pi)$. Hence, for all $\theta \in [0, 2\pi)$,

$$G(e^{j\theta}) + G^*(e^{j\theta})$$
$$= D + D^T + C\Psi(j\theta)^{-1}B + B^T\Psi(-j\theta)^{-T}C^T$$
$$\geq V + \left(C - B^T P A\right)\Psi(j\theta)^{-1}B + B^T\Psi(-j\theta)^{-T}\left(C^T - A^T P B\right)$$
$$\quad + B^T\Psi(-j\theta)^{-T}H\Psi(j\theta)^{-1}B$$
$$= V + \left[\left(C - B^T P A\right)H^{-1} + B^T\Psi(-j\theta)^{-T}\right]$$
$$\quad \times H\left[H^{-1}\left(C^T - A^T P B\right) + \Psi(j\theta)^{-1}B\right]$$
$$\quad - \left(C - B^T P A\right)H^{-1}\left(C^T - A^T P B\right)$$
$$\geq V - \left(C - B^T P A\right)H^{-1}\left(C^T - A^T P B\right). \tag{7.38}$$

Note that

$$H - \left(C^T - A^T P B\right)V^{-1}\left(C - B^T P A\right) = Q_1 > 0.$$

This, by Schur complement, implies

$$\begin{bmatrix} H & \left(C^T - A^T P B\right) \\ \left(C - B^T P A\right) & V \end{bmatrix} > 0.$$

Therefore,

$$V - \left(C - B^T P A\right)H^{-1}\left(C^T - A^T P B\right) > 0.$$

Then, it follows from this and (7.38) that (7.35) is satisfied for all $\theta \in [0, 2\pi)$. Finally, from the inequalities in (7.28) and (7.35), and by Definition 7.2, it can be easily seen that the discrete singular system in (7.23) and (7.24) is ESPR.

(*Necessity*) Suppose that the discrete singular system in (7.23) and (7.24) is admissible and ESPR. Then, there exist nonsingular matrices M and N such that (7.29) and (7.30) hold. In this case the transfer function is given in (7.31). Thus, by Lemma 7.3, we have that there exists a matrix $\tilde{P}_1 > 0$ such that

$$\begin{bmatrix} A_1^T \tilde{P}_1 A_1 - \tilde{P}_1 & C_1^T - A_1^T \tilde{P}_1 B_1 \\ C_1 - B_1^T \tilde{P}_1 A_1 & \Upsilon \end{bmatrix} < 0, \tag{7.39}$$

where

$$\Upsilon = C_2 B_2 + B_2^T C_2^T - \left(D^T + D\right) + B_1^T \tilde{P}_1 B_1.$$

It then follows from (7.39) that there exists a matrix $\tilde{P}_2 < 0$ such that

$$\begin{bmatrix} A_1^T \tilde{P}_1 A_1 - \tilde{P}_1 & C_1^T - A_1^T \tilde{P}_1 B_1 & 0 \\ C_1 - B_1^T \tilde{P}_1 A_1 & \Upsilon & C_2 \\ 0 & C_2^T & \tilde{P}_2 \end{bmatrix} < 0. \tag{7.40}$$

Pre- and post-multiplying (7.40) by

$$\begin{bmatrix} I & 0 & 0 \\ 0 & 0 & I \\ 0 & I & -B_2^T \end{bmatrix},$$

and its transpose, respectively, we obtain

$$
\begin{bmatrix}
A_1^T \tilde{P}_1 A_1 - \tilde{P}_1 & 0 & C_1^T - A_1^T \tilde{P}_1 B_1 \\
0 & \tilde{P}_2 & C_2^T - \tilde{P}_2 B_2 \\
C_1 - B_1^T \tilde{P}_1 A_1 & C_2 - B_2^T \tilde{P}_2 & B_1^T \tilde{P}_1 B_1 + B_2^T \tilde{P}_2 B_2 - (D^T + D)
\end{bmatrix} < 0.
$$

This, by the notations in (7.29) and (7.30), can be rewritten as

$$
\begin{bmatrix}
(MAN)^T \tilde{P} MAN - (MEN)^T \tilde{P} MEN \\
CN - (MB)^T \tilde{P} MAN
\end{bmatrix}
$$

$$
\left.
\begin{matrix}
N^T C^T - (MAN)^T \tilde{P} MB \\
-\left(D^T + D - (MB)^T \tilde{P} MB\right)
\end{matrix}
\right] < 0, \tag{7.41}
$$

where

$$
\tilde{P} = \begin{bmatrix} \tilde{P}_1 & 0 \\ 0 & \tilde{P}_2 \end{bmatrix}.
$$

Pre- and post-multiplying (7.41) by $\mathrm{diag}\left(N^{-T}, I\right)$ and its transpose, respectively, we obtain

$$
\begin{bmatrix}
A^T M^T \tilde{P} MA - E^T M^T \tilde{P} ME & C^T - A^T M^T \tilde{P} MB \\
C - B^T M^T \tilde{P} MA & -(D^T + D - B^T M^T \tilde{P} MB)
\end{bmatrix} < 0. \tag{7.42}
$$

Now, set

$$
P = M^T \tilde{P} M. \tag{7.43}
$$

Then, by (7.42), it is easy to see that the matrix P given in (7.43) satisfies (7.26) and (7.27). This completes the proof. $\qquad\square$

Remark 7.2. In the case when $E = I$, it is easy to see that the positive real lemma in Theorem 7.4 reduces to that in Lemma 7.3 for state-space systems. Therefore, Theorem 7.4 can be viewed as an extension of the positive real lemma for discrete state-space systems to discrete singular systems. $\qquad\triangleleft$

It is noted that the version of the positive real lemma in Theorem 7.4 involves a non-strict LMI (7.26). We now provide another version in terms of a strict LMI, which is given in the following theorem.

Theorem 7.5. *The discrete singular system in (7.23) and (7.24) is admissible and ESPR if and only if there exist matrices $P > 0$ and Q such that the following LMI holds:*

$$
\begin{bmatrix}
A^T PA - E^T PE & C^T - A^T PB \\
C - B^T PA & -(D^T + D - B^T PB)
\end{bmatrix}
$$

$$
+ \begin{bmatrix} A^T \\ -B^T \end{bmatrix} SQ^T + QS^T \begin{bmatrix} A^T \\ -B^T \end{bmatrix}^T < 0, \tag{7.44}
$$

where $S \in \mathbb{R}^{n \times (n-r)}$ is any matrix with full column rank and satisfies $E^T S = 0$.

Proof. (Sufficiency) Assume that the LMI in (7.44) is satisfied. We first show the admissibility of the system (7.23). To this end, we write

$$Q = \begin{bmatrix} Q_1 \\ Q_2 \end{bmatrix},$$

where $Q_1 \in \mathbb{R}^{n \times (n-r)}$ and $Q_2 \in \mathbb{R}^{l \times (n-r)}$. Then, the LMI in (7.44) can be rewritten as

$$\begin{bmatrix} A^T P A - E^T P E + A^T S Q_1^T + Q_1 S^T A \\ C - B^T P A + Q_2 S^T A - B^T S Q_1^T \end{bmatrix.$$

$$\left. \begin{matrix} C^T - A^T P B + A^T S Q_2^T - Q_1 S^T B \\ -(D^T + D - B^T P B + B^T S Q_2^T + Q_2 S^T B) \end{matrix} \right] < 0, \tag{7.45}$$

which implies

$$A^T P A - E^T P E + A^T S Q_1^T + Q_1 S^T A < 0.$$

Therefore, by Theorem 2.4, we have that system (7.23) is admissible. Next, we show that under the condition of the theorem, the discrete singular system in (7.23) and (7.24) is ESPR. To this end, we apply Schur complement to (7.45) and obtain

$$A^T P A - E^T P E + A^T S Q_1^T + Q_1 S^T A + \Omega < 0, \tag{7.46}$$

where

$$\Omega = \Xi^T U^{-1} \Xi, \tag{7.47}$$
$$U = D^T + D - B^T P B + B^T S Q_2^T + Q_2 S^T B > 0, \tag{7.48}$$
$$\Xi = C - B^T P A + Q_2 S^T A - B^T S Q_1^T. \tag{7.49}$$

Then, it is easy to see that (7.46) implies that there exists a matrix $W > 0$ such that

$$A^T P A - E^T P E + A^T S Q_1^T + Q_1 S^T A + \Omega + W < 0. \tag{7.50}$$

Let

$$\Phi(j\theta) = e^{j\theta} E - A.$$

Then, recalling that system (7.23) is admissible, we have that $\Phi(j\theta)$ is non-singular for all $\theta \in [0, 2\pi)$. Pre- and post-multiplying (7.50) by $B^T \Phi(-j\theta)^{-T}$ and $\Phi(j\theta)^{-1} B$, respectively, yield

$$-B^T \Phi(-j\theta)^{-T} \left[A^T P A - E^T P E + A^T S Q_1^T + Q_1 S^T A \right] \Phi(j\theta)^{-1} B$$
$$\geq B^T \Phi(-j\theta)^{-T} (\Omega + W) \Phi(j\theta)^{-1} B. \tag{7.51}$$

Now, by some straightforward algebraic manipulations, it can be verified that

$$
\begin{aligned}
& B^T P B + B^T \Phi(-j\theta)^{-T} \left(A^T P + Q_1 S^T \right) B \\
& + B^T \left(A^T P + Q_1 S^T \right)^T \Phi(j\theta)^{-1} B \\
& = -B^T \Phi(-j\theta)^{-T} \left(A^T P A - E^T P E \right. \\
& \left. + A^T S Q_1^T + Q_1 S^T A \right) \Phi(j\theta)^{-1} B,
\end{aligned}
\tag{7.52}
$$

where the relationship $E^T S = 0$ is used. From (7.51) and (7.52), it is easy to show that for all $\theta \in [0, 2\pi)$,

$$
\begin{aligned}
& B^T P B + B^T \Phi(-j\theta)^{-T} \left(A^T P + Q_1 S^T \right) B \\
& + B^T \left(A^T P + Q_1 S^T \right)^T \Phi(j\theta)^{-1} B \\
& \geq B^T \Phi(-j\theta)^{-T} \left(\Omega + W \right) \Phi(j\theta)^{-1} B.
\end{aligned}
\tag{7.53}
$$

Note that for all $\theta \in [0, 2\pi)$,

$$
\begin{aligned}
& G(e^{j\theta}) + G(e^{j\theta})^* \\
& = D + D^T + C\Phi(j\theta)^{-1} B + \left(C\Phi(j\theta)^{-1} B \right)^* \\
& = U + B^T P B + \Xi\Phi(j\theta)^{-1} B + B^T \Phi(-j\theta)^{-T} \Xi^T \\
& \quad + \left[\left(B^T P A - Q_2 S^T A + B^T S Q_1^T \right) \Phi(j\theta)^{-1} - Q_2 S^T \right] B \\
& \quad + B^T \left[\Phi(-j\theta)^{-T} \left(A^T P B - A^T S Q_2^T + Q_1 S^T B \right) - S Q_2^T \right] \\
& = U + \Xi\Phi(j\theta)^{-1} B + B^T \Phi(-j\theta)^{-T} \Xi^T \\
& \quad + B^T P B + B^T \Phi(-j\theta)^{-T} \left(A^T P + Q_1 S^T \right) B \\
& \quad + B^T \left(A^T P + Q_1 S^T \right)^T \Phi(j\theta)^{-1} B,
\end{aligned}
\tag{7.54}
$$

where U is given in (7.48). Then, by (7.53) and (7.54), we have that for all $\theta \in [0, 2\pi)$,

$$
\begin{aligned}
G(e^{j\theta}) + G(e^{j\theta})^* \geq {} & U + \Xi\Phi(j\theta)^{-1} B + B^T \Phi(-j\theta)^{-T} \Xi^T \\
& + B^T \Phi(-j\theta)^{-T} \left(\Omega + W \right) \Phi(j\theta)^{-1} B.
\end{aligned}
$$

Noting $\Omega + W > 0$ and using Lemma 4.6, we obtain

$$
G(e^{j\theta}) + G(e^{j\theta})^* \geq U - \Xi \left(\Omega + W \right)^{-1} \Xi^T.
\tag{7.55}
$$

Observe that

$$
\Omega + W - \Xi^T U^{-1} \Xi = W > 0,
$$

which, by Schur complement, implies

$$
\begin{bmatrix} U & \Xi \\ \Xi^T & \Omega + W \end{bmatrix} > 0.
\tag{7.56}
$$

By Schur complement again, it follows from (7.56) that

$$U - \Xi \left(\Omega + W\right)^{-1} \Xi^T > 0. \tag{7.57}$$

Therefore, from (7.55) and (7.57), we have that for all $\theta \in [0, 2\pi)$,

$$G(e^{j\theta}) + G^*(e^{j\theta}) > 0. \tag{7.58}$$

On the other hand, since the discrete singular system in (7.23) and (7.24) is admissible, we can find two nonsingular matrices $\tilde{M}_1$ and $\tilde{N}_1$ such that

$$\tilde{M}_1 E \tilde{N}_1 = \begin{bmatrix} I & 0 \\ 0 & 0 \end{bmatrix}, \quad \tilde{M}_1 A \tilde{N}_1 = \begin{bmatrix} \tilde{A}_1 & 0 \\ 0 & I \end{bmatrix}. \tag{7.59}$$

In this case, the matrix S satisfying $E^T S = 0$ can be chosen as

$$S = \tilde{M}_1^T \begin{bmatrix} 0 \\ I \end{bmatrix} \tilde{H}_1,$$

where $\tilde{H}_1 \in \mathbb{R}^{(n-r) \times (n-r)}$ is any nonsingular matrix. Now, write

$$C \tilde{N}_1 = \begin{bmatrix} \tilde{C}_1 & \tilde{C}_2 \end{bmatrix}, \quad \tilde{M}_1 B = \begin{bmatrix} \tilde{B}_1 \\ \tilde{B}_2 \end{bmatrix},$$

$$\tilde{M}_1^{-T} P \tilde{M}_1^{-1} = \begin{bmatrix} \tilde{P}_1 & \tilde{P}_2 \\ \tilde{P}_2^T & \tilde{P}_3 \end{bmatrix}, \quad \tilde{N}_1^T Q_1 \tilde{H}_1^T = \begin{bmatrix} \tilde{Q}_{11} \\ \tilde{Q}_{21} \end{bmatrix},$$

$$Q_2 \tilde{H}_1^T = \tilde{Q}_2,$$

where the partition is compatible with that of $\tilde{M}_1 E \tilde{N}_1$ and $\tilde{M}_1 A \tilde{N}_1$ in (7.59). Then, it is easy to see that

$$\tilde{P}_1 > 0. \tag{7.60}$$

Pre- and post-multiplying (7.45) by $\mathrm{diag}(\tilde{N}_1^T, I)$ and $\mathrm{diag}(\tilde{N}_1, I)$, respectively and then using the above notations, we have

$$\begin{bmatrix} * & * & * \\ * & \tilde{P}_3 + \tilde{Q}_{21} + \tilde{Q}_{21}^T & J \\ * & J^T & \mathcal{V} \end{bmatrix} < 0. \tag{7.61}$$

where $*$ represents a matrix that will not be used in the following discussion, and

$$J = \tilde{C}_2^T - \tilde{P}_2^T \tilde{B}_1 - \tilde{P}_3 \tilde{B}_2 + \tilde{Q}_2^T - \tilde{Q}_{21} \tilde{B}_2,$$
$$\mathcal{V} = -D^T - D + \tilde{B}_1^T \tilde{P}_1 \tilde{B}_1 + \tilde{B}_2^T \tilde{P}_2^T \tilde{B}_1$$
$$\quad + \tilde{B}_1^T \tilde{P}_2 \tilde{B}_2 + \tilde{B}_2^T \tilde{P}_3 \tilde{B}_2 - \tilde{B}_2^T \tilde{Q}_2^T - \tilde{Q}_2 \tilde{B}_2.$$

Pre- and post-multiplying (7.61) by

$$\begin{bmatrix} I & 0 & 0 \\ 0 & I & 0 \\ 0 & \tilde{B}_2^T & I \end{bmatrix},$$

and its transpose, respectively, and then noting the 3-3 block, we obtain

$$-D^T - D + \tilde{C}_2\tilde{B}_2 + \tilde{C}_2^T\tilde{B}_2^T + \tilde{B}_1^T\tilde{P}_1\tilde{B}_1 < 0.$$

With this and (7.60), we have

$$D - \tilde{C}_2\tilde{B}_2 + D^T + \tilde{C}_2^T\tilde{B}_2^T > 0,$$

which together with

$$G(z) = C(zE - A)^{-1}B = \tilde{C}_1(zI - \tilde{A}_1)^{-1}\tilde{B}_1 - \tilde{C}_2\tilde{B}_2 + D$$

implies

$$G(\infty) + G(\infty)^T > 0.$$

Considering this and (7.58), it follows easily from Definition 7.2 that the discrete singular system in (7.23) and (7.24) is ESPR.

(*Necessity*) Suppose that the discrete singular system in (7.23) and (7.24) is admissible and ESPR. Then, there exist two nonsingular matrices $\mathcal{M}$ and $\mathcal{N}$ such that

$$E = \mathcal{M}\begin{bmatrix} I & 0 \\ 0 & 0 \end{bmatrix}\mathcal{N}, \quad A = \mathcal{M}\begin{bmatrix} \mathcal{A} & 0 \\ 0 & I \end{bmatrix}\mathcal{N}. \tag{7.62}$$

Then, it is easy to see that the matrix S satisfying $E^T S = 0$ can be written as

$$S = \mathcal{M}^{-T}\begin{bmatrix} 0 \\ I \end{bmatrix}\mathcal{H}, \tag{7.63}$$

where $\mathcal{H} \in \mathbb{R}^{(n-r) \times (n-r)}$ is any nonsingular matrix. Write

$$B = \mathcal{M}\begin{bmatrix} B_1 \\ B_2 \end{bmatrix}, \quad C = \begin{bmatrix} C_1 & C_2 \end{bmatrix}\mathcal{N},$$

where the partition is compatible with that of A. Note that

$$G(z) = C_1(zI - \mathcal{A})^{-1}B_1 + D - C_2B_2.$$

Considering the discrete singular system in (7.23) and (7.24) is ESPR, by Lemma 7.3, it follows that there exists a matrix $\tilde{P} > 0$ such that

$$\begin{bmatrix} \mathcal{A}^T\tilde{P}\mathcal{A} - \tilde{P} & C_1^T - \mathcal{A}^T\tilde{P}B_1 \\ C_1 - B_1^T\tilde{P}\mathcal{A} & -\left(D^T + D - C_2B_2 - B_2^TC_2^T - B_1^T\tilde{P}B_1\right) \end{bmatrix} < 0. \tag{7.64}$$

Let

$$Q_1 = 0, \quad Q_2 = -I, \quad Q_3 = -C_2 + B_2^T.$$

Then, by (7.64), it is easy to see that

$$
\left[
\begin{array}{cc}
\mathcal{A}^T \tilde{P} \mathcal{A} - \tilde{P} & C_1^T - \mathcal{A}^T \tilde{P} B_1 \\
C_1 - B_1^T \tilde{P} \mathcal{A} - \left(D^T + D - C_2 B_2 - B_2^T C_2^T - B_1^T \tilde{P} B_1 \right) \\
Q_1^T & C_2^T + Q_3^T + Q_2^T B_2
\end{array}
\right.
$$

$$
\left.
\begin{array}{c}
Q_1 \\
C_2 + Q_3 + B_2^T Q_2 \\
I + Q_2 + Q_2^T
\end{array}
\right] < 0.
$$

Pre- and post-multiplying this inequality by

$$
\begin{bmatrix}
I & 0 & 0 \\
0 & 0 & I \\
0 & I & -B_2^T
\end{bmatrix},
$$

and its transpose, respectively, we obtain

$$
\left[
\begin{array}{cc|c}
\mathcal{A}^T \tilde{P} \mathcal{A} - \tilde{P} & Q_1 & \Lambda_1 \\
Q_1^T & I + Q_2 + Q_2^T & \Lambda_2 \\
\hline
\Lambda_1^T & \Lambda_2^T & \Lambda_3
\end{array}
\right] < 0, \tag{7.65}
$$

where

$$
\begin{aligned}
\Lambda_1 &= C_1^T - \mathcal{A}^T \tilde{P} B_1 - Q_1 B_2, \\
\Lambda_2 &= C_2^T - B_2 + Q_3^T - Q_2 B_2, \\
\Lambda_3 &= -D^T - D + B_1^T \tilde{P} B_1 + B_2^T B_2 - B_2^T Q_3^T - Q_3 B_2.
\end{aligned}
$$

Set

$$
\mathcal{P} = \mathcal{M}^{-T} \begin{bmatrix} \tilde{P} & 0 \\ 0 & I \end{bmatrix} \mathcal{M}, \quad
\mathcal{Q} = \begin{bmatrix} \mathcal{N} & 0 \\ \hline 0 & I \end{bmatrix}^T \begin{bmatrix} Q_1 \\ Q_2 \\ Q_3 \end{bmatrix} \mathcal{H}^{-T}. \tag{7.66}
$$

Then, pre- and post-multiplying (7.65) by

$$
\begin{bmatrix} \mathcal{N} & 0 \\ \hline 0 & I \end{bmatrix}^T
$$

and its transpose, respectively, we can write

$$
\begin{bmatrix}
A^T \mathcal{P} A - E^T \mathcal{P} E & C^T - A^T \mathcal{P} B \\
C - B^T \mathcal{P} A & -\left(D^T + D - B^T \mathcal{P} B \right)
\end{bmatrix}
$$

$$
+ \begin{bmatrix} A^T \\ -B^T \end{bmatrix} S \mathcal{Q}^T + \mathcal{Q} S^T \begin{bmatrix} A^T \\ -B^T \end{bmatrix}^T < 0. \tag{7.67}
$$

That is, the matrices $\mathcal{P}$ and $\mathcal{Q}$ given in (7.66) satisfy (7.44). This completes the proof. $\square$

Having the positive real lemma in hand, we now study the positive real control problem for discrete singular systems.

Consider the following state feedback controller

$$u(k) = Kx(k), \quad K \in \mathbb{R}^{m \times n}. \tag{7.68}$$

Applying this controller to the singular system in (7.21) and (7.22) results in the following closed-loop system:

$$Ex(k+1) = (A + B_1 K) x(k) + Bw(k), \tag{7.69}$$
$$z(k) = Cx(k) + Dw(k). \tag{7.70}$$

Then, the positive real control problem to be addressed is to design a state feedback controller (7.68) such that the closed-loop system in (7.69) and (7.70) is admissible and ESPR.

A necessary and sufficient condition for the solvability of the positive real control problem is presented in the following theorem.

Theorem 7.6. *Given the discrete singular system in (7.21) and (7.22). There exists a state feedback controller in (7.68) such that the closed-loop system in (7.69) and (7.70) is admissible and ESPR if and only if there exist a scalar $\delta > 0$, matrices $P > 0$, Q_1 and Q_2 such that*

$$\mathcal{V} = D^T + D - B^T PB + Q_2 S^T B + B^T SQ_2 > 0, \tag{7.71}$$

and

$$\mathcal{W} - \mathcal{U}^T \mathcal{T}^{-1} \mathcal{U} < 0, \tag{7.72}$$

where $S \in \mathbb{R}^{n \times (n-r)}$ is any matrix with full column rank and satisfies $E^T S = 0$, and

$$
\begin{aligned}
\mathcal{W} &= Q_1 S^T A + A^T SQ_1^T - E^T PE + A^T PA \\
&\quad + \left[C^T - Q_1 S^T B + A^T \left(SQ_2^T - PB \right) \right] \\
&\quad \times \mathcal{V}^{-1} \left[C - B^T SQ_1^T + \left(Q_2 S^T - B^T P \right) A \right], \\
\mathcal{U} &= B_1^T SQ_1^T + B_1^T PA + B_1^T \left(SQ_2^T - PB \right) \mathcal{V}^{-1} \\
&\quad \times \left[C - B^T SQ_1^T + \left(Q_2 S^T - B^T P \right) A \right], \\
\mathcal{T} &= B_1^T \left[P + \left(SQ_2^T - PB \right) \mathcal{V}^{-1} \left(Q_2 S^T - B^T P \right) \right] B_1 + \delta I.
\end{aligned}
$$

In this case, a desired state feedback control law can be chosen by

$$u(k) = -\mathcal{T}^{-1} \mathcal{U} x(k). \tag{7.73}$$

Proof. By Theorem 7.5, we have that there exists a state feedback controller in (7.68) such that the closed-loop system in (7.69) and (7.70) is admissible and ESPR if and only if there exist matrices $P > 0$ and Q such that

$$\begin{bmatrix} A_c^T P A_c - E^T P E & C^T - A_c^T P B \\ C - B^T P A_c & -(D^T + D - B^T P B) \end{bmatrix}$$

$$+ \begin{bmatrix} A_c^T \\ -B^T \end{bmatrix} S Q^T + Q S^T \begin{bmatrix} A_c^T \\ -B^T \end{bmatrix}^T < 0, \tag{7.74}$$

where

$$A_c = A + B_1 K.$$

Write

$$Q = \begin{bmatrix} Q_1 \\ Q_2 \end{bmatrix},$$

where the partition is compatible with the related matrices. Then, (7.74) can be rewritten as

$$\begin{bmatrix} A_c^T P A_c + Q_1 S^T A_c + A_c^T S Q_1^T - E^T P E \\ C - (B^T P - Q_2 S^T) A_c - B^T S Q_1^T \\ \end{bmatrix}$$
$$\begin{matrix} C^T - A_c^T (P B - S Q_2^T) - Q_1 S^T B \\ -\mathcal{V} \end{matrix} \Bigg] < 0.$$

which, by Schur complement, is equivalent to $\mathcal{V} > 0$, and

$$\mathcal{W} + K^T \mathcal{U} + \mathcal{U}^T K$$
$$+ K^T B_1^T \left[P + (S Q_2^T - P B) \mathcal{V}^{-1} (Q_2 S^T - B^T P) \right] B_1 K < 0. \tag{7.75}$$

It is easy to see that (7.75) is satisfied if and only if there exists a scalar $\delta > 0$ such that

$$\mathcal{W} + K^T \mathcal{U} + \mathcal{U}^T K + K^T \mathcal{T} K < 0.$$

This can be rewritten as

$$\mathcal{W} - \mathcal{U}^T \mathcal{T}^{-1} \mathcal{U} + \left[K^T + \mathcal{U}^T \mathcal{T}^{-1} \right] \mathcal{T} \left[K + \mathcal{T}^{-1} \mathcal{U} \right] < 0. \tag{7.76}$$

Finally, it is easy to see that there exists a matrix K such that (7.76) holds if and only if (7.72) holds; in this case, a desired K can be chosen as in (7.73). This completes the proof. $\qquad\square$

7.4 Conclusion

This chapter has addressed the positive real control problem for singular systems. Versions of positive real lemma have been obtained for both continuous and discrete singular systems. Based on these results, necessary and sufficient conditions, which guarantee the existence of a state feedback controller such that the closed-loop system is admissible and extended strictly positive real, have been proposed. In the context of continuous singular systems, a desired state feedback controller can be constructed by solving a strict LMI, while in the context of discrete singular system, a desired state feedback controller can be constructed by solving a matrix inequality. In both the continuous and discrete cases, the proposed results can be viewed as extensions of positive real control for state-space systems to singular systems. Part of the results presented in this chapter have appeared in [178, 203].

8

H_∞ Filtering

8.1 Introduction

The H_∞ filtering problem for state-space systems has received much attention in the past two decades. In the H_∞ setting, the noises are assumed to be arbitrary deterministic signals with bounded energy (or average power). Compared with the traditional Kalman filtering, the H_∞ filtering approach does not require knowledge of the statistical properties of the external noises, and the designed filter in the H_∞ setting is more robust against additional parameter uncertainties in the system model. These properties make the H_∞ filtering technique useful in many applications. A great number of results on H_∞ filtering for state-space systems have been reported in the literature; see, for example, [60, 130, 142, 193], and the references cited therein.

This chapter is concerned with the problem of H_∞ filtering for singular systems. First, the full-order H_∞ filtering problem for continuous singular systems is addressed. The purpose is to design a linear proper filter such that the resulting error system is regular, impulse-free and stable while the closed-loop transfer function from the disturbance to the filtering error output satisfies a prescribed H_∞-norm bound constraint. A necessary and sufficient condition for the solvability of this problem is obtained in terms of LMIs. The desired filters can be chosen with McMillan degree no more than the number of the exponential modes of the plant. Second, the problem of reduced-order H_∞ filtering is considered in both the contexts of continuous and discrete singular systems. Attention is focused on the design of a linear proper filter, having a specified order lower than the order of the system under consideration, such that the filtering error dynamic system is regular, impulse-free (for the continuous case) or causal (for the discrete case) and stable, while the closed-loop transfer function satisfies a prescribed H_∞ performance level. Necessary and sufficient conditions for the solvability of the reduced-order H_∞ problem are obtained in terms of certain LMIs and a coupling non-convex rank constraint set. Furthermore, it is shown that the solvability of the zeroth-order H_∞ filtering problem is equivalent to a convex LMI feasibility problem. An explicit parametrization of all desired reduced-order filters is also provided. The derived results in this chapter can be viewed as extensions of existing results on H_∞ filtering for state-space systems to singular systems.

8.2 Full-Order Filtering

In this section, we are concerned with the full-order H_∞ filtering problem for continuous singular systems. The class of linear singular systems to be considered is described by

$$E\dot{x}(t) = Ax(t) + B\omega(t), \tag{8.1}$$
$$y(t) = Cx(t), \tag{8.2}$$
$$z(t) = Lx(t), \tag{8.3}$$

where $x(t) \in \mathbb{R}^n$ is the state; $y(t) \in \mathbb{R}^m$ is the measurement; $z(t) \in \mathbb{R}^q$ is the signal to be estimated; $\omega(t) \in \mathbb{R}^p$ is the disturbance input which belongs to $\mathcal{L}_2[0, \infty)$. The matrix $E \in \mathbb{R}^{n \times n}$ may be singular; we shall assume that rank$(E) = r \leq n$. A, B, C and L are known real constant matrices with appropriate dimensions.

Now, we consider the following filter for the estimate of $z(t)$:

$$E_f \dot{\hat{x}}(t) = A_f \hat{x}(t) + B_f y(t), \tag{8.4}$$
$$\hat{z}(t) = C_f \hat{x}(t) + D_f y(t), \tag{8.5}$$

where $\hat{x}(t) \in \mathbb{R}^{\hat{n}}$ and $\hat{z}(t) \in \mathbb{R}^q$ are the state and the output of the filter, respectively. The matrices $E_f \in \mathbb{R}^{\hat{n} \times \hat{n}}$, $A_f \in \mathbb{R}^{\hat{n} \times \hat{n}}$, $B_f \in \mathbb{R}^{\hat{n} \times p}$, $C_f \in \mathbb{R}^{q \times \hat{n}}$ and $D_f \in \mathbb{R}^{q \times m}$ are to be determined. Let

$$e(t) = \begin{bmatrix} x(t)^T & \hat{x}(t)^T \end{bmatrix}^T, \quad \tilde{z}(t) = z(t) - \hat{z}(t). \tag{8.6}$$

Then the filtering error dynamics from the system in (8.1)–(8.3) and the filter in (8.4) and (8.5) can be written as

$$E_c \dot{e}(t) = A_c e(t) + B_c \omega(t), \tag{8.7}$$
$$\tilde{z}(t) = L_c e(t), \tag{8.8}$$

where

$$E_c = \begin{bmatrix} E & 0 \\ 0 & E_f \end{bmatrix}, \quad A_c = \begin{bmatrix} A & 0 \\ B_f C & A_f \end{bmatrix}, \tag{8.9}$$

$$B_c = \begin{bmatrix} B \\ 0 \end{bmatrix}, \quad L_c = \begin{bmatrix} L - D_f C & -C_f \end{bmatrix}. \tag{8.10}$$

The H_∞ filter problem to be addressed can be formulated as follows: given the singular system in (8.1)–(8.3) and a prescribed H_∞ bound $\gamma > 0$, determine a filter in the form of (8.4) and (8.5) such that the filtering error system in (8.7) and (8.8) is admissible and the transfer function from $w(t)$ to $\tilde{z}(t)$ given as

$$G_c(s) = L_c(sE_c - A_c)^{-1}B_c, \tag{8.11}$$

satisfies

$$\|G_c\|_\infty < \gamma. \tag{8.12}$$

Remark 8.1. In this section, we do not put any constraint on the order of the filter to be designed. ◁

We present a necessary and sufficient condition for the solvability of the H_∞ filtering problem for continuous singular systems in the following theorem.

Theorem 8.1. *Given a linear continuous singular system in (8.1)–(8.3) and a scalar $\gamma > 0$. There exists a filter in the form of (8.4) and (8.5) such that the H_∞ filtering problem is solvable if and only if there exist matrices X, Y, Φ, Ψ, Ξ and Υ such that the following LMIs hold:*

$$E^T X = X^T E \geq 0, \tag{8.13}$$
$$E^T Y = Y^T E \geq 0, \tag{8.14}$$
$$E^T (X - Y) \geq 0, \tag{8.15}$$

$$\begin{bmatrix} A^T Y + Y^T A & \Theta_1 & Y^T B & L^T - \Upsilon^T - C^T \Xi^T \\ \Theta_1^T & \Theta_2 & X^T B & L^T - C^T \Xi^T \\ B^T Y & B^T X & -\gamma^2 I & 0 \\ L - \Upsilon - \Xi C & L - \Xi C & 0 & -I \end{bmatrix} < 0, \tag{8.16}$$

where

$$\Theta_1 = A^T X + Y^T A + C^T \Psi^T + \Phi^T,$$
$$\Theta_2 = X^T A + A^T X + \Psi C + C^T \Psi^T.$$

In this case, there exist nonsingular matrices S, $\tilde{S}$, W and $\tilde{W}$ such that

$$E^T \tilde{S} = S^T E, \tag{8.17}$$
$$EW = \tilde{W}^T E^T, \tag{8.18}$$
$$XY^{-1} = I - \tilde{S}W, \tag{8.19}$$
$$Y^{-1}X = I - \tilde{W}S. \tag{8.20}$$

Then, the parameters of a desired H_∞ filter given in the form of (8.4) and (8.5) can be chosen as

$$E_f = E, \quad A_f = S^{-T}\Phi Y^{-1}W^{-1}, \tag{8.21}$$
$$B_f = S^{-T}\Psi, \quad C_f = \Upsilon Y^{-1}W^{-1}, \quad D_f = \Xi. \tag{8.22}$$

Proof. (*Sufficiency*) Suppose that the LMIs in (8.13)–(8.16) hold. Under these conditions, we first show that there always exist nonsingular matrices S, $\tilde{S}$, W and $\tilde{W}$ such that (8.17)–(8.20) hold. To this end, we claim that the matrix Y satisfying (8.14)–(8.16) is nonsingular. If not, then there will exist a vector $\eta \neq 0$ such that $Y\eta = 0$. Thus, $\eta^T\left(A^TY + Y^TA\right)\eta = 0$, which is a contradiction since (8.16) gives $A^TY + Y^TA < 0$. Also, we assume that, without loss of generality, $Y - X$ is nonsingular. If not, we can choose $\tilde{Y} = (1 - \alpha)Y$, where $\alpha > 0$ is a sufficiently small scalar which is not an eigenvalue of $I - XY^{-1}$ and satisfies that (8.16) holds for $\tilde{Y}$. Then, it can be seen that (8.14) and (8.15) are satisfied for this $\tilde{Y}$. Furthermore, $\tilde{Y} - X$ is nonsingular. Therefore, we can replace Y by this $\tilde{Y}$ to satisfy the above requirements without violating (8.14)–(8.16). Now, it is easy to see that both $I - XY^{-1}$ and $I - Y^{-1}X$ are nonsingular. Choose two nonsingular matrices M and N such that

$$E = M \begin{bmatrix} I & 0 \\ 0 & 0 \end{bmatrix} N.$$

Considering this and (8.13) and (8.14), we can deduce that the matrices X and Y can be written as

$$X = M^{-T} \begin{bmatrix} X_1 & 0 \\ X_2 & X_3 \end{bmatrix} N, \quad Y = M^{-T} \begin{bmatrix} Y_1 & 0 \\ Y_2 & Y_3 \end{bmatrix} N, \tag{8.23}$$

where

$$X_1^T = X_1 \geq 0, \quad Y_1^T = Y_1 > 0.$$

Then

$$Y^{-1} = N^{-1} \begin{bmatrix} \hat{Y}_1 & 0 \\ \hat{Y}_2 & \hat{Y}_3 \end{bmatrix} M^T,$$

where

$$\hat{Y}_1^T = \hat{Y}_1 = Y_1^{-1} > 0.$$

Set

$$S = M^{-T} \begin{bmatrix} S_1 & 0 \\ S_2 & S_3 \end{bmatrix} N, \quad \tilde{S} = M^{-T} \begin{bmatrix} \tilde{S}_1 & 0 \\ \tilde{S}_2 & \tilde{S}_3 \end{bmatrix} N, \tag{8.24}$$

$$W = N^{-1} \begin{bmatrix} W_1 & 0 \\ W_2 & W_3 \end{bmatrix} M^T, \quad \tilde{W} = N^{-1} \begin{bmatrix} \tilde{W}_1 & 0 \\ \tilde{W}_2 & \tilde{W}_3 \end{bmatrix} M^T, \tag{8.25}$$

where the matrices S_i, $\tilde{S}_i$, W_i, $\tilde{W}_i$, $i = 1, 2, 3$, are selected to satisfy

$$S_1^T = \tilde{S}_1, \qquad W_1 = \tilde{W}_1^T, \tag{8.26}$$

$$\begin{bmatrix} \tilde{S}_1 W_1 & 0 \\ \tilde{S}_2 W_1 + \tilde{S}_3 W_2 & \tilde{S}_3 W_3 \end{bmatrix} = \begin{bmatrix} I - X_1 \hat{Y}_1 & 0 \\ -X_2 \hat{Y}_1 - X_3 \hat{Y}_2 & I - X_3 \hat{Y}_3 \end{bmatrix}, \tag{8.27}$$

$$\begin{bmatrix} \tilde{W}_1 S_1 & 0 \\ \tilde{W}_2 S_1 + \tilde{W}_3 S_2 & \tilde{W}_3 S_3 \end{bmatrix} = \begin{bmatrix} I - \hat{Y}_1 X_1 & 0 \\ -\hat{Y}_2 X_1 - \hat{Y}_3 X_3 & I - \hat{Y}_3 X_3 \end{bmatrix}. \tag{8.28}$$

By (8.23)–(8.28), it can be verified that the matrices S, $\tilde{S}$, W and $\tilde{W}$ given in (8.24) and (8.25) satisfy (8.17)–(8.20). Furthermore, the nonsingularity of $I - XY^{-1}$ and $I - Y^{-1}X$ implies that the matrices S, $\tilde{S}$, W and $\tilde{W}$ are also nonsingular.

Next, we will show that the error system in (8.7) and (8.8), derived from the singular system in (8.1)–(8.3) and the filter in (8.4) and (8.5) whose parameters are given in (8.21) and (8.22), is admissible and its transfer function satisfies (8.12). For this purpose, we define

$$\Lambda_1 = \begin{bmatrix} \bar{Y} & I \\ W & 0 \end{bmatrix}, \qquad \Lambda_2 = \begin{bmatrix} I & X \\ 0 & S \end{bmatrix},$$

where $\bar{Y} = Y^{-1}$. It is noted that both Λ_1 and Λ_2 are nonsingular. Set

$$\hat{P} = \Lambda_2 \Lambda_1^{-1}.$$

Then it can be verified that $\hat{P}$ is nonsingular, and

$$\hat{P} = \begin{bmatrix} X & \tilde{S} \\ S & -\Gamma \end{bmatrix},$$

where

$$\Gamma = S\bar{Y}W^{-1}. \tag{8.29}$$

It follows from (8.17)–(8.20) that

$$\begin{aligned} E^T \Gamma &= W^{-T} \left(W^T E^T S\bar{Y} \right) W^{-1} \\ &= W^{-T} E\tilde{W} S\bar{Y}W^{-1} \\ &= W^{-T} E \left(I - \bar{Y}X \right) \bar{Y}W^{-1} \\ &= W^{-T} E \left(\bar{Y} - \bar{Y}X\bar{Y} \right) W^{-1} \\ &= W^{-T} E\bar{Y} \left(Y - X \right) \bar{Y}W^{-1} \\ &= W^{-T} \bar{Y}^T \left[E^T \left(Y - X \right) \right] \bar{Y}W^{-1}. \end{aligned}$$

By (8.13)–(8.15), we have

$$E^T \Gamma = \Gamma^T E \le 0. \tag{8.30}$$

Therefore,

$$\hat{E}^T \hat{P} = \hat{P}^T \hat{E}, \tag{8.31}$$

where $\hat{E} = \mathrm{diag}(E, E)$. Noting the nonsingularity of Γ and using (8.30), (8.17) and (8.19), we have

$$
\begin{aligned}
&E^T X + E^T \tilde{S} \Gamma^{-1} \left(\Gamma^{-T} E^T \right)^+ \Gamma^{-T} E^T S \\
&= E^T X + S^T E \Gamma^{-1} \left(\Gamma^{-T} E^T \right)^+ \Gamma^{-T} E^T S \\
&= E^T X + S^T \Gamma^{-T} E^T \left(\Gamma^{-T} E^T \right)^+ \Gamma^{-T} E^T S \\
&= E^T X + S^T \Gamma^{-T} E^T S \\
&= E^T X + S^T E \Gamma^{-1} S \\
&= E^T X + E^T \tilde{S} \Gamma^{-1} S \\
&= E^T (X + \tilde{S} \Gamma^{-1} S) \\
&= E^T (X + \tilde{S} W Y) \\
&= E^T \left[X + \left(I - X \bar{Y} \right) Y \right] \\
&= E^T Y \\
&\geq 0.
\end{aligned} \tag{8.32}
$$

Furthermore, considering that $E \Gamma^{-1}$ is symmetric, we obtain

$$
\begin{aligned}
&E^T \tilde{S} \Gamma^{-1} \left[I - \left(-\Gamma^{-T} E^T \right) \left(-\Gamma^{-T} E^T \right)^+ \right] \\
&= S^T E \Gamma^{-1} \left[I - (E \Gamma^{-1})^T \left(\left((E \Gamma^{-1})^+ \right)^T \right) \right] \\
&= S^T E \Gamma^{-1} \left[I - \left((E \Gamma^{-1})^+ (E \Gamma^{-1}) \right)^T \right] \\
&= S^T E \Gamma^{-1} \left[I - (E \Gamma^{-1})^+ (E \Gamma^{-1}) \right] \\
&= S^T \left[E \Gamma^{-1} - (E \Gamma^{-1}) (E \Gamma^{-1})^+ (E \Gamma^{-1}) \right] \\
&= 0.
\end{aligned} \tag{8.33}
$$

Taking into account (8.32) and (8.33) and using Lemma 3.1, we can deduce

$$
\begin{bmatrix} E^T X & E^T \tilde{S} \Gamma^{-1} \\ \Gamma^{-T} E^T S & -\Gamma^{-T} E^T \end{bmatrix} \geq 0. \tag{8.34}
$$

Pre- and post-multiplying (8.34) by $\mathrm{diag}(I, \Gamma^T)$ and its transpose, respectively, give

$$
\begin{bmatrix} E^T X & E^T \tilde{S} \\ E^T S & -E^T \Gamma \end{bmatrix} \geq 0. \tag{8.35}
$$

By noting (8.31), we can rewrite (8.35) as

$$\hat{E}^T \hat{P} = \hat{P}^T \hat{E} \geq 0. \tag{8.36}$$

On the other hand, pre- and post-multiplying (8.16) by $\mathrm{diag}(\bar{Y}^T, I, I, I)$ and its transpose, respectively, we have

$$
\begin{bmatrix}
\bar{Y}^T A^T + A\bar{Y} & \bar{Y}^T \Theta_1 & B & \bar{Y}^T L^T - \bar{Y}^T \Upsilon^T - \bar{Y}^T C^T \Xi^T \\
\Theta_1^T \bar{Y} & \Theta_2 & X^T B & L^T - C^T \Xi^T \\
B^T & B^T X & -\gamma^2 I & 0 \\
L\bar{Y} - \Upsilon \bar{Y} - \Xi C\bar{Y} & L - \Xi C & 0 & -I
\end{bmatrix} < 0.
$$

This can be rewritten as

$$
\begin{bmatrix}
\Lambda_1^T \left(A_c^T \hat{P} + \hat{P}^T A_c \right) \Lambda_1 & \Lambda_1^T \hat{P}^T B_c & \Lambda_1^T L_c^T \\
B_c^T \hat{P} \Lambda_1 & -\gamma^2 I & 0 \\
L_c \Lambda_1 & 0 & -I
\end{bmatrix} < 0, \tag{8.37}
$$

where the matrices A_c, B_c and L_c are given in (8.9) and (8.10) with the parameters A_f, B_f, C_f and D_f given in (8.21) and (8.22). Then, pre- and post-multiplying (8.37) by $\mathrm{diag}(\Lambda_1^{-T}, I, I)$ and its transpose, respectively, we obtain

$$
\begin{bmatrix}
A_c^T \hat{P} + \hat{P}^T A_c & \hat{P}^T B_c & L_c^T \\
B_c^T \hat{P} & -\gamma^2 I & 0 \\
L_c & 0 & -I
\end{bmatrix} < 0. \tag{8.38}
$$

Noting this and (8.36) and using Theorem 5.1, we have that the error system in (8.7) and (8.8) resulting from the singular system in (8.1)–(8.3) and the filter in (8.4) and (8.5) with the parameters given in (8.21) and (8.22) is admissible and its transfer function satisfies (8.12). This completes the proof of sufficiency.

(*Necessity*) Suppose that the H_∞ filtering problem is solvable; that is, there exists a filter in the form of (8.4) and (8.5) such that the error system in (8.7) and (8.8) is admissible and its transfer function satisfies (8.12). Then, by Theorem 5.1, it can be seen that there exists a matrix P_c such that

$$
E_c^T P_c = P_c^T E_c \geq 0, \tag{8.39}
$$

$$
\begin{bmatrix}
A_c^T P_c + P_c^T A_c & P_c^T B_c & L_c^T \\
B_c^T P_c & -\gamma^2 I & 0 \\
L_c & 0 & -I
\end{bmatrix} < 0. \tag{8.40}
$$

Without loss of generality, we assume

$$
\hat{n} \geq n + \mathrm{rank}(E_f). \tag{8.41}
$$

If not, we can choose $\tilde{E}_f \in \mathbb{R}^{\tilde{n} \times \tilde{n}}$, $\tilde{A}_f \in \mathbb{R}^{\tilde{n} \times \tilde{n}}$, $\tilde{B}_f \in \mathbb{R}^{\tilde{n} \times p}$ and $\tilde{C}_f \in \mathbb{R}^{q \times \tilde{n}}$ to replace E_f, A_f, B_f and C_f, respectively, such that $\tilde{n} \geq n + \mathrm{rank}(E_f)$ and (8.39) and (8.40) still hold for some $\tilde{P}_c$ as follows. First, set

$$\tilde{E}_f = \begin{bmatrix} E_f & 0 \\ 0 & 0 \end{bmatrix}, \quad \tilde{A}_f = \begin{bmatrix} A_f & 0 \\ 0 & -I \end{bmatrix}, \tag{8.42}$$

$$\tilde{B}_f = \begin{bmatrix} B_f \\ 0 \end{bmatrix}, \quad \tilde{C}_f = \begin{bmatrix} C_f & 0 \end{bmatrix}, \tag{8.43}$$

with $\tilde{n} \geq n + \mathrm{rank}(E_f)$. Then, choose

$$\tilde{P}_c = \begin{bmatrix} P_c & 0 \\ 0 & I \end{bmatrix}. \tag{8.44}$$

It can be verified that $\tilde{E}_f$, $\tilde{A}_f$, $\tilde{B}_f$ and $\tilde{C}_f$ in (8.42) and (8.43) together with $\tilde{P}_c$ in (8.44) satisfy (8.39) and (8.40). Therefore, the required assumption is satisfied. Now, it is easy to see that the LMI in (8.40) implies

$$A_c^T P_c + P_c^T A_c < 0. \tag{8.45}$$

This implies that the matrix P_c is nonsingular. Write P_c and P_c^{-1} as

$$P_c = \begin{bmatrix} P_{c1} & P_{c2} \\ P_{c3} & P_{c4} \end{bmatrix}, \quad P_c^{-1} = \begin{bmatrix} Q_{c1} & Q_{c2} \\ Q_{c3} & Q_{c4} \end{bmatrix}. \tag{8.46}$$

where the partition is compatible with that of A_c. Next, we will show that P_{c4} and Q_{c1} are nonsingular. To this end, we note from (8.39) that

$$E^T P_{c1} = P_{c1}^T E \geq 0, \tag{8.47}$$
$$E_f^T P_{c4} = P_{c4}^T E_f \geq 0. \tag{8.48}$$

On the other hand, it is easy to see that the 2-2 block of (8.45) gives

$$P_{c4}^T A_f + A_f^T P_{c4} < 0, \tag{8.49}$$

which implies that P_{c4} is nonsingular. Now, pre- and post-multiplying (8.39) by P_c^{-T} and its transpose, respectively, result in

$$P_c^{-T} E_c^T = E_c P_c^{-1} \geq 0.$$

The 1-1 block of this inequality gives

$$Q_{c1}^T E^T = E Q_{c1} \geq 0. \tag{8.50}$$

Pre- and post-multiplying (8.45) by P_c^{-T} and its transpose, respectively, we have

$$P_c^{-T} A_c^T + A_c P_c^{-1} < 0.$$

Then the 1-1 block of this inequality gives

$$A Q_{c1} + Q_{c1}^T A < 0.$$

This implies that Q_{c1} is nonsingular. In the following, without loss of generality, we assume that the matrix P_{c3} is of full column rank. If not, we can choose another $\hat{P}_c$ to replace P_c such that $\hat{P}_c$ satisfies the assumption and the inequalities (8.39) and (8.40) simultaneously as follows. First, considering (8.41), we choose a matrix $\Omega \in \mathbb{R}^{\hat{n} \times n}$ with full column rank satisfying $E_f^T \Omega = 0$. Then, we choose a nonsingular matrix Q such that

$$Q\Omega = \begin{bmatrix} \Omega_1 \\ 0 \end{bmatrix},$$

where $\Omega_1 \in \mathbb{R}^{n \times n}$ is nonsingular. Partition QP_{c3} as

$$QP_{c3} = \begin{bmatrix} P_{c31} \\ P_{c32} \end{bmatrix}$$

where $P_{c31} \in \mathbb{R}^{n \times n}$. Now, we choose a sufficiently small scalar $\alpha > 0$ such that α is not an eigenvalue of $-\Omega_1^{-1} P_{c31}$ and the matrix $\hat{P}_c$ defined as

$$\hat{P}_c = P_c + \alpha \begin{bmatrix} 0 & 0 \\ \Omega & 0 \end{bmatrix} \tag{8.51}$$

satisfies (8.39) and (8.40). Then, the 2-1 block of $\hat{P}_c$ in (8.51) is

$$P_{c3} + \alpha \Omega = Q^{-1} (QP_{c3} + \alpha Q\Omega) = Q^{-1} \begin{bmatrix} \alpha\Omega_1 + P_{c31} \\ P_{c32} \end{bmatrix},$$

which is of full column rank. Note that

$$E_c^T \hat{P}_c = \hat{P}_c^T E_c = E_c^T P_c \geq 0.$$

Thus, $\hat{P}_c$ given in (8.51) satisfies the assumption. Now, it can be verified that

$$Q_{c3} = -P_{c4}^{-1} P_{c3} (P_{c1} - P_{c2} P_{c4}^{-1} P_{c3})^{-1}.$$

Since P_{c3} can be chosen to be of full column rank, we have that Q_{c3} can also be chosen to be of full column rank. Set

$$\Pi_1 = \begin{bmatrix} Q_{c1} & I \\ Q_{c3} & 0 \end{bmatrix}, \qquad \Pi_2 = \begin{bmatrix} I & P_{c1} \\ 0 & P_{c3} \end{bmatrix}.$$

Then, it can be seen that both Π_1 and Π_2 are of full column rank. Furthermore, it is easy to see that

$$P_c \Pi_1 = \Pi_2. \tag{8.52}$$

Now, pre- and post-multiplying (8.40) by $\mathrm{diag}(\Pi_1^T, I, I)$ and its transpose, respectively, we have

$$\begin{bmatrix} AQ_{c1} + Q_{c1}^T A^T \\ A^T + P_{c1}^T AQ_{c1} + P_{c3}^T B_f CQ_{c1} + P_{c3}^T A_f Q_{c3} \\ B^T \\ LQ_{c1} - C_f Q_{c3} - D_f CQ_{c1} \\ \\ A + Q_{c1}^T A^T P_{c1} + Q_{c1}^T C^T B_f^T P_{c3} + Q_{c3}^T A_f^T P_{c3} \\ P_{c1}^T A + A^T P_{c1} + P_{c3}^T B_f C + C^T B_f^T P_{c3} \\ B^T P_{c1} + D^T B_f^T P_{c3} \\ L - D_f C \\ \\ B \qquad Q_{c1}^T L^T - Q_{c3}^T C_f^T - Q_{c1}^T C^T D_f^T \\ P_{c1}^T B + P_{c3}^T B_f D \qquad L^T - C^T D_f^T \\ -\gamma^2 I \qquad 0 \\ 0 \qquad -I \end{bmatrix} < 0, \qquad (8.53)$$

where the relationship in (8.52) is used. Pre- and post-multiplying (8.53) by $\mathrm{diag}(Q_{c1}^{-T}, I, I, I)$ and its transpose, respectively, give

$$\begin{bmatrix} Q_{c1}^{-T} A + A^T Q_{c1}^{-1} \\ A^T Q_{c1}^{-1} + P_{c1}^T A + P_{c3}^T B_f C + P_{c3}^T A_f Q_{c3} Q_{c1}^{-1} \\ B^T Q_{c1}^{-1} \\ L - C_f Q_{c3} Q_{c1}^{-1} - D_f C \\ \\ Q_{c1}^{-T} A + A^T P_{c1} + C^T B_f^T P_{c3} + Q_{c1}^{-T} Q_{c3}^T A_f^T P_{c3} \\ P_{c1}^T A + A^T P_{c1} + P_{c3}^T B_f C + C^T B_f^T P_{c3} \\ B^T P_{c1} + D^T B_f^T P_{c3} \\ L - D_f C \\ \\ Q_{c1}^{-T} B \qquad L^T - Q_{c1}^{-T} Q_{c3}^T C_f^T - C^T D_f^T \\ P_{c1}^T B + P_{c3}^T B_f D \qquad L^T - C^T D_f^T \\ -\gamma^2 I \qquad 0 \\ 0 \qquad -I \end{bmatrix} < 0. \qquad (8.54)$$

Set

$$\begin{aligned} X &= P_{c1}, & Y &= Q_{c1}^{-1}, \\ \varPhi &= P_{c3}^T A_f Q_{c3} Q_{c1}^{-1}, & \varPsi &= P_{c3}^T B_f, \\ \varUpsilon &= C_f Q_{c3} Q_{c1}^{-1}, & \varXi &= D_f. \end{aligned}$$

Then, the inequalities in (8.47) and (8.54) provide (8.13) and (8.16), respectively. Also, (8.50) implies

$$E^T Q_{c1}^{-1} = Q_{c1}^{-T} E \geq 0,$$

which gives (8.14). Note that

$$Q_{c1} = \left(P_{c1} - P_{c2} P_{c4}^{-1} P_{c3} \right)^{-1}.$$

Therefore, by (8.48), we have

$$E^T(P_{c1} - Q_{c1}^{-1}) = E^T P_{c2} P_{c4}^{-1} P_{c3} = P_{c3}^T E_f P_{c4}^{-1} P_{c3} \geq 0,$$

which provides (8.15). This completes the proof. $\qquad\square$

Remark 8.2. Theorem 8.1 provides a necessary and sufficient condition for the design of filters for continuous singular systems such that the resulting error system is admissible and the transfer function satisfies a prescribed H_∞-norm bound constraint. The optimal filter in the sense that the minimum H_∞ norm γ is approached can be obtained by solving the following optimization problem:

$$\min_{X,Y,\Phi,\Psi,\varUpsilon} \gamma$$

subject to $\gamma > 0$ and the LMIs in (8.13)–(8.16). $\qquad\lhd$

Remark 8.3. It can be seen that Theorem 8.1 shows that the designed filter in (8.4) and (8.5) with parameters given by (8.21) and (8.22) is admissible and the number of the finite modes of this filter equals to rank(E); that is, the designed filter is proper with a McMillan degree no more than the number of the exponential modes of the given singular system. $\qquad\lhd$

Remark 8.4. In the case when the matrix $E = I$, that is, the singular system in (8.1)–(8.3) reduces to a state-space system, it can be shown that the result presented in Theorem 8.1 coincides with that in [86] if there are no parameter uncertainties in the state-space model. Therefore, Theorem 8.1 can be regarded as an extension of the existing H_∞ results for state-space systems to singular systems. $\qquad\lhd$

8.3 Reduced-Order Filtering

In this section, the problem of reduced-order H_∞ filtering problem is considered for both continuous and discrete singular systems. The designed filters are with a specified order lower than that of the system under consideration.

8.3.1 Continuous Systems

Consider a linear continuous singular system in (8.1)–(8.3). For this system, we consider the following filter:

$$\dot{\xi}(t) = \hat{A}\xi(t) + \hat{B}y(t), \tag{8.55}$$

$$\hat{z}(t) = \hat{C}\xi(t) + \hat{D}y(t), \tag{8.56}$$

where $\xi(t) \in \mathbb{R}^{\hat{n}}$, $\hat{z}(t) \in \mathbb{R}^q$, and $\hat{n} < r$; the matrices $\hat{A} \in \mathbb{R}^{\hat{n} \times \hat{n}}$, $\hat{B} \in \mathbb{R}^{\hat{n} \times m}$, $\hat{C} \in \mathbb{R}^{q \times \hat{n}}$ and $\hat{D} \in \mathbb{R}^{q \times m}$ are to be determined. Let

$$\eta(t) = \begin{bmatrix} x(t)^T & \xi(t)^T \end{bmatrix}^T, \quad \tilde{z}(t) = z(t) - \hat{z}(t). \tag{8.57}$$

Then, the filtering error dynamics from the system in (8.1)–(8.3) and the filter in (8.55) and (8.56) can be written as:

$$\tilde{E}\dot{\eta}(t) = \tilde{A}\eta(t) + \tilde{B}\omega(t), \tag{8.58}$$

$$\tilde{z}(t) = \tilde{C}\eta(t), \tag{8.59}$$

where

$$\tilde{E} = \begin{bmatrix} E & 0 \\ 0 & I \end{bmatrix}, \quad \tilde{A} = \begin{bmatrix} A & 0 \\ \hat{B}C & \hat{A} \end{bmatrix}, \tag{8.60}$$

$$\tilde{B} = \begin{bmatrix} B \\ 0 \end{bmatrix}, \quad \tilde{C} = \begin{bmatrix} L - \hat{D}C & -\hat{C} \end{bmatrix}. \tag{8.61}$$

The reduced-order H_∞ filtering problem to be addressed can be formulated as follows: given the continuous singular system in (8.1)–(8.3) and a prescribed H_∞-norm bound $\gamma > 0$, determine a filter in the form of (8.55) and (8.56) such that

(RC1) the error dynamic system in (8.58) and (8.59) is admissible.

(RC2) the transfer function $\hat{G}_{ce}(s)$ of the error dynamic system satisfies

$$\|\tilde{G}_{ce}\|_\infty < \gamma, \tag{8.62}$$

where

$$\tilde{G}_{ce}(s) = \tilde{C}(s\tilde{E} - \tilde{A})^{-1}\tilde{B}.$$

In the case when $\hat{n} = 0$, the reduced-order H_∞ filter in (8.55) and (8.56) is specialized as

$$\hat{z}(t) = \hat{D}y(t). \tag{8.63}$$

In this case, the reduced-order H_∞ filtering problem reduces to the static or zeroth-order H_∞ filtering problem [66].

Before proceeding further, we give the following lemma which will be used in the development. The proof can be carried out by using Theorem 5.1, and is thus omitted.

Lemma 8.1. *Given a scalar $\gamma > 0$. The following statements (S1) and (S2) are equivalent.*

(S1) The following conditions (i) and (ii) hold simultaneously.

(i) The singular system in (8.1) under $\omega(t) \equiv 0$ is admissible.

(ii) The transfer function $G(s)$ of the singular system in (8.1) and (8.2) given as

$$G(s) = C(sE - A)^{-1}B,$$

satisfies

$$\|G\|_\infty < \gamma.$$

(S2) There exists a matrix P satisfying the following LMIs:

$$E^T P = P^T E \geq 0,$$

$$\begin{bmatrix} A^T P + P^T A & P^T B & C^T \\ B^T P & -\gamma I & 0 \\ C & 0 & -\gamma I \end{bmatrix} < 0.$$

Now we present a necessary and sufficient condition for the solvability of the reduced-order H_∞ filtering problem for continuous singular systems.

Theorem 8.2. *There exists an $\hat{n}$-th order filter in the form of (8.55) and (8.56) to solve the reduced-order H_∞ filtering problem for the continuous singular system in (8.1)–(8.3) if and only if there exist matrices $X \geq 0$ and Y satisfying*

$$\begin{bmatrix} Y^T A + A^T Y & Y^T B \\ B^T Y & -\gamma I \end{bmatrix} < 0, \tag{8.64}$$

$$\begin{bmatrix} \vartheta & \mathcal{N}(XE + Y)^T B & \mathcal{N}L^T \\ B^T(XE + Y)\mathcal{N}^T & -\gamma I & 0 \\ L\mathcal{N}^T & 0 & -\gamma I \end{bmatrix} < 0, \tag{8.65}$$

$$E^T Y = Y^T E \geq 0, \tag{8.66}$$

and

$$\text{rank}(X) \leq \hat{n}, \tag{8.67}$$

where

$$\vartheta = \mathcal{N}\left[(XE + Y)^T A + A^T(XE + Y)\right]\mathcal{N}^T, \tag{8.68}$$

$$\mathcal{N} = \left(C^T\right)^\perp. \tag{8.69}$$

In this case, the parameters of all desired filters with order $\hat{n}$ corresponding to a feasible solution (X, Y) to (8.64)–(8.67) are given by

$$\begin{bmatrix} \hat{D} & \hat{C} \\ \hat{B} & \hat{A} \end{bmatrix} = \left[-\hat{W}^{-1} \Psi^T \Lambda \Phi_r^T (\Phi_r \Lambda \Phi_r^T)^{-1} \right.$$
$$\left. + \hat{W}^{-1} \Xi^{\frac{1}{2}} \Upsilon (\Phi_r \Lambda \Phi_r^T)^{-\frac{1}{2}} \right] \Phi_l^+ + \Theta - \Theta \Phi_l \Phi_l^+, \qquad (8.70)$$

where

$$\Xi = \hat{W} - \Psi^T \left[\Lambda - \Lambda \Phi_r^T \left(\Phi_r \Lambda \Phi_r^T \right)^{-1} \Phi_r \Lambda \right] \Psi,$$

$$\Lambda = (\Psi \hat{W}^{-1} \Psi^T - \Omega)^{-1},$$

$$\Omega = \begin{bmatrix} \chi & A^T X_{12} & (XE+Y)^T B & L^T \\ X_{12}^T A & 0 & X_{12}^T B & 0 \\ B^T (XE+Y) & B^T X_{12} & -\gamma I & 0 \\ L & 0 & 0 & -\gamma I \end{bmatrix},$$

$$\Psi = \begin{bmatrix} 0 & E^T X_{12} \\ 0 & X_{22} \\ 0 & 0 \\ -I & 0 \end{bmatrix},$$

$$\Phi = \begin{bmatrix} C & 0 & 0 & 0 \\ 0 & I & 0 & 0 \end{bmatrix},$$

$$\chi = (XE+Y)^T A + A^T (XE+Y),$$

and Θ is any matrix with appropriate dimensions; Υ is any matrix satisfying $\|\Upsilon\| < 1$; Φ_l and Φ_r are any full rank factors of Φ; that is, $\Phi = \Phi_l \Phi_r$; moreover, $X_{12} \in \mathbb{R}^{n \times \hat{n}}$, $X_{22} \in \mathbb{R}^{\hat{n} \times \hat{n}}$, $X_{22} > 0$ and $\hat{W} > 0$ satisfying $\Lambda > 0$ and

$$X = X_{12} X_{22}^{-1} X_{12}^T \geq 0.$$

Proof. Note that the matrices $\tilde{A}$ and $\tilde{C}$ in (8.60) and (8.61) can be rewritten as

$$\tilde{A} = \bar{A} + \bar{F} G \bar{H}, \quad \tilde{C} = \bar{C} + \bar{S} G \bar{H}, \qquad (8.71)$$

with

$$\bar{A} = \begin{bmatrix} A & 0 \\ 0 & 0 \end{bmatrix}, \quad \bar{C} = \begin{bmatrix} L & 0 \end{bmatrix}, \qquad (8.72)$$

$$G = \begin{bmatrix} \hat{D} & \hat{C} \\ \hat{B} & \hat{A} \end{bmatrix}, \quad \bar{F} = \begin{bmatrix} 0 & 0 \\ 0 & I \end{bmatrix}, \qquad (8.73)$$

$$\bar{H} = \begin{bmatrix} C & 0 \\ 0 & I \end{bmatrix}, \quad \bar{S} = \begin{bmatrix} -I & 0 \end{bmatrix}. \qquad (8.74)$$

Using Lemma 8.1, we have that the filtering error system in (8.58) and (8.59) satisfies the requirements (RC1) and (RC2) if and only if there exists a matrix $\tilde{P} > 0$ such that

$$\tilde{E}_c^T \tilde{P} = \tilde{P}^T \tilde{E}_c \geq 0, \tag{8.75}$$

$$\begin{bmatrix} \tilde{A}^T \tilde{P} + \tilde{P}^T \tilde{A} & \tilde{P}^T \tilde{B} & \tilde{C}^T \\ \tilde{B}^T \tilde{P} & -\gamma I & 0 \\ \tilde{C} & 0 & -\gamma I \end{bmatrix} < 0. \tag{8.76}$$

By (8.76), it is easy to see that

$$\tilde{A}^T \tilde{P} + \tilde{P}^T \tilde{A} < 0, \tag{8.77}$$

which implies that the matrix $\tilde{P}$ is nonsingular. Now, write $\tilde{P}$ and $\tilde{P}^{-1}$ as

$$\tilde{P} = \begin{bmatrix} W & W_1 \\ W_2 & W_3 \end{bmatrix}, \quad \tilde{P}^{-1} = \begin{bmatrix} Z & Z_1 \\ Z_2 & Z_3 \end{bmatrix}, \tag{8.78}$$

where the partition is compatible with $\bar{A}$ in (8.72). Then, from (8.75), we have

$$\begin{bmatrix} E^T W & E^T W_1 \\ W_2 & W_3 \end{bmatrix} = \begin{bmatrix} W^T E & W_2^T \\ W_1^T E & W_3^T \end{bmatrix} \geq 0, \tag{8.79}$$

which implies

$$E^T W = W^T E \geq 0, \tag{8.80}$$
$$W_3 = W_3^T \geq 0, \tag{8.81}$$
$$W_2 = W_1^T E. \tag{8.82}$$

Note that (8.77) implies

$$\hat{A}^T W_3 + W_3^T \hat{A} < 0.$$

Then, using (8.81), we have $W_3 > 0$. Now, applying Schur complement to (8.79) gives

$$E^T \tilde{W} = \tilde{W}^T E \geq 0, \tag{8.83}$$

where

$$\tilde{W} = W - W_1 W_3^{-1} W_1^T E. \tag{8.84}$$

Since $\tilde{P}$ is nonsingular, the LMIs in (8.75) and (8.77) can be rewritten as

$$\tilde{P}^{-T} \tilde{E}_c^T = \tilde{E}_c \tilde{P}^{-1} \geq 0,$$
$$\tilde{P}^{-T} \tilde{A}^T + \tilde{A} \tilde{P}^{-1} < 0.$$

Substituting $\tilde{P}^{-1}$ in (8.78) into these two inequalities, we obtain

$$Z^T E^T = EZ \geq 0,$$
$$Z^T A^T + AZ < 0.$$

These two inequalities imply that the matrix Z is nonsingular. Then, by noting (8.78) and manipulating, we have

$$Z^{-1} = \tilde{W}. \tag{8.85}$$

On the other hand, substituting the expressions in (8.71) into the LMI in (8.76) gives

$$\begin{bmatrix} (\bar{A} + \bar{F}G\bar{H})^T \tilde{P} + \tilde{P}^T (\bar{A} + \bar{F}G\bar{H}) & \tilde{P}^T \tilde{B} & (\bar{C} + \bar{S}G\bar{H})^T \\ \tilde{B}^T \tilde{P} & -\gamma I & 0 \\ \bar{C} + \bar{S}G\bar{H} & 0 & -\gamma I \end{bmatrix} < 0,$$

which can be rewritten as

$$\Omega_c + \Psi_c G \Phi_c + (\Psi_c G \Phi_c)^T < 0, \tag{8.86}$$

where

$$\Omega_c = \begin{bmatrix} \bar{A}^T \tilde{P} + \tilde{P}^T \bar{A} & \tilde{P}^T \tilde{B} & \bar{C}^T \\ \tilde{B}^T \tilde{P} & -\gamma I & 0 \\ \bar{C} & 0 & -\gamma I \end{bmatrix},$$

$$\Psi_c = \begin{bmatrix} \tilde{P}^T \bar{F} \\ 0 \\ \bar{S} \end{bmatrix},$$

$$\Phi_c = \begin{bmatrix} \bar{H} & 0 & 0 \end{bmatrix}.$$

By Lemma 7.2, a necessary and sufficient condition for the LMI in (8.86) to have a solution G is that the following two inequalities hold simultaneously:

$$\Psi_c^\perp \Omega_c \Psi_c^{\perp T} < 0, \tag{8.87}$$

$$\Phi_c^{T\perp} \Omega_c \Phi_c^{T\perp T} < 0. \tag{8.88}$$

Observe that

$$\Psi_c^\perp = \begin{bmatrix} [I\ 0]\ 0\ 0 \\ [0\ 0]\ I\ 0 \end{bmatrix} \begin{bmatrix} \tilde{P}^{-T} & 0 & 0 \\ 0 & I & 0 \\ 0 & 0 & I \end{bmatrix},$$

$$\Phi_c^{T\perp} = \begin{bmatrix} [\mathcal{N}\ 0]\ 0\ 0 \\ [0\ 0]\ I\ 0 \\ [0\ 0]\ 0\ I \end{bmatrix}.$$

Then (8.87) and (8.88) become

$$\begin{bmatrix} AZ + Z^T A^T & B \\ B^T & -\gamma I \end{bmatrix} < 0, \tag{8.89}$$

$$\begin{bmatrix} \mathcal{N}(W^T A + A^T W)\mathcal{N}^T & \mathcal{N}W^T B & \mathcal{N}L^T \\ B^T W \mathcal{N}^T & -\gamma I & 0 \\ L \mathcal{N}^T & 0 & -\gamma I \end{bmatrix} < 0. \tag{8.90}$$

Set

$$X = W_1 W_3^{-1} W_1^T, \tag{8.91}$$

$$Y = Z^{-1}. \tag{8.92}$$

Then, considering $W_3 > 0$ and noting (8.84) and (8.85), we have $X \geq 0$ and

$$W = XE + Y. \tag{8.93}$$

Now, substituting (8.92) and (8.93) into (8.89) and (8.90), we can derive (8.64) and (8.65), respectively. Note that $W_3 \in \mathbb{R}^{\hat{n} \times \hat{n}}$ and $W_3 > 0$. Then, by the expression in (8.91), the inequality in (8.67) follows immediately, while the inequality in (8.66) can be obtained by noting (8.83), (8.85) and (8.92). Finally, when the inequalities (8.64)–(8.67) are feasible, the parametrization of all $\hat{n}$-th order filters corresponding to a feasible solution can be obtained by using the results in [55] and [84]. This completes the proof. $\qquad\square$

Remark 8.5. Theorem 8.2 provides a necessary and sufficient condition for the solvability of the reduced-order H_∞ filtering problem for continuous singular systems. It should be pointed out that the inequalities in (8.64)–(8.67) are non-convex although the constraints in (8.64)–(8.66) are convex. Fortunately, to solve these non-convex inequalities, we can resort to an efficient numerical algorithm based on alternating projections given in [65]. $\qquad\triangleleft$

In the case when $E = I$, that is, the continuous singular system in (8.1)–(8.3) reduces to a state-space one, we have the following reduced-order H_∞ filtering result.

Corollary 8.1. *Consider the continuous state-space system described by*

$$\dot{x}(t) = Ax(t) + Bw(t), \tag{8.94}$$

$$y(t) = Cx(t), \tag{8.95}$$

$$z(t) = Lx(t). \tag{8.96}$$

There exists an $\hat{n}$-th order filter in the form of (8.55) and (8.56) to solve the reduced-order H_∞ filtering problem for this system if and only if there exist matrices $X \geq 0$ and $Y > 0$ satisfying

$$\begin{bmatrix} YA + A^T Y & YB \\ B^T Y & -\gamma I \end{bmatrix} < 0, \tag{8.97}$$

$$\begin{bmatrix} \tilde{\vartheta} & \mathcal{N}(X+Y)B & \mathcal{N}L^T \\ B^T(X+Y)\mathcal{N}^T & -\gamma I & 0 \\ L\mathcal{N}^T & 0 & -\gamma I \end{bmatrix} < 0, \tag{8.98}$$

and

$$\text{rank}(X) \leq \hat{n}, \tag{8.99}$$

where

$$\tilde{\vartheta} = \mathcal{N}\left[(X+Y)A + A^T(X+Y)\right]\mathcal{N}^T,$$

and $\mathcal{N}$ is given in (8.69).

Remark 8.6. To obtain another equivalent form of Corollary 8.1, we note that (8.98) is equivalent to

$$\begin{bmatrix} \gamma\tilde{\vartheta} & \gamma\mathcal{N}(X+Y)B & \gamma\mathcal{N}L^T \\ \gamma B^T(X+Y)\mathcal{N}^T & -\gamma^2 I & 0 \\ \gamma L\mathcal{N}^T & 0 & -\gamma^2 I \end{bmatrix} < 0. \tag{8.100}$$

Pre- and post-multiply (8.100) by $\text{diag}(I, I, \gamma^{-1}I)$, and then using Schur complement, we have

$$\begin{bmatrix} \mathcal{N}\left[\gamma(X+Y)A + \gamma A^T(X+Y) + L^TL\right]\mathcal{N}^T & \gamma\mathcal{N}(X+Y)B \\ \gamma B^T(X+Y)\mathcal{N}^T & -\gamma^2 I \end{bmatrix} < 0. \tag{8.101}$$

Note that (8.97) is equivalent to

$$\begin{bmatrix} \gamma YA + \gamma A^T Y & \gamma YB \\ \gamma B^T Y & -\gamma^2 I \end{bmatrix} < 0. \tag{8.102}$$

Then by setting

$$\bar{X} = \gamma Y, \quad \bar{Y} = \gamma X + \bar{X},$$

and considering (8.101) and (8.102), we have an equivalent form of Corollary 8.1 as follows:

There exists an $\hat{n}$-th order filter in the form of (8.55) and (8.56) to solve the reduced-order H_∞ filtering problem for the state-space system in (8.94)–(8.96) if and only if there exist matrices $\bar{X} > 0$ and $\bar{Y} > 0$ satisfying

$$\begin{bmatrix} \bar{X}A + A^T\bar{X} & \bar{X}B \\ B^T\bar{X} & -\gamma^2 I \end{bmatrix} < 0, \tag{8.103}$$

$$\begin{bmatrix} \mathcal{N}\left(\bar{Y}A + A^T\bar{Y} + L^TL\right)\mathcal{N}^T & \mathcal{N}\bar{Y}B \\ B^T\bar{Y}\mathcal{N}^T & -\gamma^2 I \end{bmatrix} < 0, \tag{8.104}$$

$$\text{rank}\left(\bar{X} - \bar{Y}\right) \leq \hat{n}. \tag{8.105}$$

Now, it can be seen that in the case when $D = 0$, the results on the reduced-order H_∞ filtering problem for state-space systems obtained in Theorem 1 in [66] are the same as those in (8.103)–(8.105). Therefore, Theorem 8.2 here can be viewed as an extension of the reduced-order H_∞ filtering results for continuous state-space systems to singular systems. ◁

The next theorem presents the result on the solvability of the zeroth-order H_∞ filtering problem, which will be a simple necessary and sufficient condition in LMIs. More specifically, the solution is free from the rank constraint.

Theorem 8.3. *There exists a filter in the form of (8.63) to solve the zeroth-order H_∞ filtering problem for the continuous singular system in (8.1)–(8.3) if and only if there exists a matrix $X > 0$ satisfying*

$$\begin{bmatrix} X^T A + A^T X & X^T B \\ B^T X & -\gamma I \end{bmatrix} < 0, \qquad (8.106)$$

$$\begin{bmatrix} \mathcal{N} \left(X^T A + A^T X \right) \mathcal{N}^T & \mathcal{N} X^T B & \mathcal{N} L^T \\ B^T X \mathcal{N}^T & -\gamma I & 0 \\ L \mathcal{N}^T & 0 & -\gamma I \end{bmatrix} < 0, \qquad (8.107)$$

$$E^T X = X^T E \geq 0, \qquad (8.108)$$

where $\mathcal{N}$ is given in (8.69). In this case, all the desired zeroth-order filters in (8.63) corresponding to a feasible solution X to (8.106)–(8.108) are given by

$$\hat{D} = \left[-\hat{W}^{-1} \hat{\Psi}_c^T \Lambda \hat{\Phi}_{cr}^T (\hat{\Phi}_{cr} \Lambda \hat{\Phi}_{cr}^T)^{-1} \right.$$
$$\left. + \hat{W}^{-1} \Xi^{\frac{1}{2}} \Upsilon (\hat{\Phi}_{cr} \Lambda \hat{\Phi}_{cr}^T)^{-\frac{1}{2}} \right] \hat{\Phi}_{cl}^+ + \Theta_c - \Theta_c \hat{\Phi}_{cl} \hat{\Phi}_{cl}^+,$$

where

$$\Xi = \hat{W} - \hat{\Psi}_c^T \left[\Lambda - \Lambda \hat{\Phi}_{cr}^T \left(\hat{\Phi}_{cr} \Lambda \hat{\Phi}_{cr}^T \right)^{-1} \hat{\Phi}_{cr} \Lambda \right] \hat{\Psi}_c,$$

$$\Lambda = (\hat{\Psi}_c \hat{W}^{-1} \hat{\Psi}_c^T - \hat{\Omega}_c)^{-1},$$

$$\hat{\Omega}_c = \begin{bmatrix} X^T A + A^T X & X^T B & L^T \\ B^T X & -\gamma I & 0 \\ L & 0 & -\gamma I \end{bmatrix},$$

$$\hat{\Psi}_c = \begin{bmatrix} 0 & 0 & -I \end{bmatrix}^T,$$

$$\hat{\Phi}_c = \begin{bmatrix} C & 0 & 0 \end{bmatrix},$$

and Θ_c is any matrix with appropriate dimensions; Υ is any matrix satisfying $\|\Upsilon\| < 1$ and $\hat{W} > 0$ such that $\Lambda > 0$; $\hat{\Phi}_{cl}$ and $\hat{\Phi}_{cr}$ are any full rank factors of $\hat{\Phi}_c$; that is, $\hat{\Phi}_c = \hat{\Phi}_{cl} \hat{\Phi}_{cr}$.

Proof. By (8.1)–(8.3) and (8.63), the filtering error system can be obtained as

$$E\dot{x}(t) = Ax(t) + B\omega(t),$$
$$\tilde{z}(t) = (L - \hat{D}C)x(t),$$

where $\tilde{z}(t)$ is defined in (8.57). By Lemma 8.1, it is easy to see that this system is admissible and satisfies a prescribed H_∞ performance level $\gamma > 0$ if and only if there exists a matrix X such that

$$E^T X = X^T E \geq 0, \tag{8.109}$$

$$\begin{bmatrix} A^T X + X^T A & X^T B & (L - \hat{D}C)^T \\ B^T X & -\gamma I & 0 \\ L - \hat{D}C & 0 & -\gamma I \end{bmatrix} < 0. \tag{8.110}$$

It can be shown that (8.110) can be rewritten as

$$\hat{\Omega}_c + \hat{\Psi}_c \hat{D} \hat{\Phi}_c + (\hat{\Psi}_c \hat{D} \hat{\Phi}_c)^T < 0. \tag{8.111}$$

Then, by Lemma 7.2, it can be seen that a necessary and sufficient condition for the above LMI to have a solution $\hat{D}$ is

$$\hat{\Psi}_c^\perp \hat{\Omega}_c \hat{\Psi}_c^{\perp T} < 0, \quad \hat{\Phi}_c^{T\perp} \hat{\Omega}_c \hat{\Phi}_c^{T\perp T} < 0. \tag{8.112}$$

The inequalities in (8.112) together with (8.109) provide the LMIs in (8.106)–(8.108), respectively. Finally, the parametrization of all $\hat{D}$ satisfying (8.111) can be obtained by using the results in [55] and [84]. This completes the proof.

$\square$

Remark 8.7. The optimal zeroth-order H_∞ filtering problem can be solved by finding a solution to the following optimization problem.

$$\min_{\gamma > 0,\, X} \gamma$$

subject to the constraints (8.106)–(8.108). It is worth pointing out that (8.106)–(8.108) are LMIs not only in the matrix variable X, but also in the scalar γ. Therefore, the above optimization problem is convex, which can be easily handled by using linear programming in convex optimization [15]. ◁

Based on Theorem 8.3, the zeroth-order H_∞ filtering problem for state-space systems can be obtained in the following corollary, which, by a similar argument as in Remark 8.6, can be verified to be equivalent to Theorem 3 in [66] with the case $D = 0$.

Corollary 8.2. *There exists a constant matrix $\hat{D}$ solving the zeroth-order H_∞ filtering problem for the state-space system in (8.94)–(8.96) if and only if there exists a matrix $X > 0$ such that*

$$\begin{bmatrix} XA + A^T X & XB \\ B^T X & -\gamma I \end{bmatrix} < 0,$$

$$\begin{bmatrix} \mathcal{N}\left(XA + A^T X\right)\mathcal{N}^T & \mathcal{N} XB & \mathcal{N} L^T \\ B^T X \mathcal{N}^T & -\gamma I & 0 \\ L\mathcal{N}^T & 0 & -\gamma I \end{bmatrix} < 0,$$

where $\mathcal{N}$ is given in (8.69).

8.3.2 Discrete Systems

Now, we investigate the reduced-order H_∞ filtering problem for discrete singular systems. The class of linear discrete singular systems to be considered is described by

$$Ex(k+1) = Ax(k) + B\omega(k), \tag{8.113}$$
$$y(k) = Cx(k) + D\omega(k), \tag{8.114}$$
$$z(k) = Lx(k), \tag{8.115}$$

where $x(k) \in \mathbb{R}^n$ is the state; $y(k) \in \mathbb{R}^m$ is the measurement; $z(k) \in \mathbb{R}^q$ is the signal to be estimated; $\omega(k) \in \mathbb{R}^p$ is the disturbance input which belongs to $l_2[0, \infty)$. The matrix $E \in \mathbb{R}^{n \times n}$ may be singular; we shall assume that rank$(E) = r \leq n$. A, B, C, D and L are known real constant matrices with appropriate dimensions.

Similar to the continuous case, we consider the following filter for the estimate of $z(k)$:

$$\xi(k+1) = \hat{A}\xi(k) + \hat{B}y(k), \tag{8.116}$$
$$\hat{z}(k) = \hat{C}\xi(k) + \hat{D}y(k), \tag{8.117}$$

where $\xi(k) \in \mathbb{R}^{\hat{n}}$, $\hat{z}(k) \in \mathbb{R}^q$, and $\hat{n} < r$; the matrices $\hat{A} \in \mathbb{R}^{\hat{n} \times \hat{n}}$, $\hat{B} \in \mathbb{R}^{\hat{n} \times m}$, $\hat{C} \in \mathbb{R}^{q \times \hat{n}}$ and $\hat{D} \in \mathbb{R}^{q \times m}$ are to be determined. Let

$$\eta(k) = \left[x(k)^T \ \xi(k)^T \right]^T, \quad \tilde{z}(k) = z(k) - \hat{z}(k). \tag{8.118}$$

Then, the filtering error dynamics generated from the discrete singular system in (8.113)–(8.115) and the filter in (8.116) and (8.117) can be written as

$$\tilde{E}\eta(k+1) = \tilde{A}\eta(k) + \tilde{B}\omega(k), \tag{8.119}$$
$$\tilde{z}(t) = \tilde{C}\eta(k) + \tilde{D}\omega(k), \tag{8.120}$$

where $\tilde{E}$, $\tilde{A}$ and $\tilde{C}$ are given in (8.60) and (8.61), and

$$\tilde{B} = \begin{bmatrix} B \\ \hat{B}D \end{bmatrix}, \quad \tilde{D} = -\hat{D}D. \tag{8.121}$$

The reduced-order H_∞ filtering problem to be addressed in the discrete case can be formulated as follows: given the discrete singular system in (8.113)–(8.115) and a prescribed H_∞-norm bound $\gamma > 0$, determine a filter in the form of (8.116) and (8.117) such that

(RD1) the error dynamic system in (8.119) and (8.120) is admissible.

(RD2) the transfer function $\hat{G}_d(z)$ of the error dynamic system satisfies

$$\|\tilde{G}_d\|_\infty < \gamma, \tag{8.122}$$

where

$$\tilde{G}_d(z) = \tilde{C}(z\tilde{E}_d - \tilde{A})^{-1}\tilde{B} + \tilde{D}. \tag{8.123}$$

In the case when $\hat{n} = 0$, the reduced-order H_∞ filter in (8.116) and (8.117) is specialized as

$$\hat{z}(k) = \hat{D}y(k). \tag{8.124}$$

We first provide a lemma which will be used in the sequel. The proof can be carried out by using Theorem 5.5, and is thus omitted.

Lemma 8.2. *Consider the discrete singular system in (8.113) and (8.114). Then, the following statements (S1) and (S2) are equivalent.*

(S1) The following conditions (i) and (ii) hold simultaneously.

(i) The discrete singular system in (8.113) and (8.114) under $\omega(k) \equiv 0$ is admissible.

(ii) The transfer function $G(z)$ of the singular system in (8.113) and (8.114) given as
$$G(z) = C(zE - A)^{-1}B + D,$$
satisfies
$$\|G\|_\infty < \gamma.$$

(S2) There exists a matrix $P = P^T$ satisfying the following LMIs:

$$E^T P E \geq 0,$$

$$\begin{bmatrix} A^T P A - E^T P E & A^T P B & C^T \\ B^T P A & B^T P B - \gamma I & D^T \\ C & D & -\gamma I \end{bmatrix} < 0.$$

Now, we present a solvability condition for the reduced-order H_∞ filtering problem for discrete singular systems.

Theorem 8.4. *There exists an $\hat{n}$-th order filter in the form of (8.116) and (8.117) to solve the reduced-order H_∞ filtering problem for the discrete singular system in (8.113)–(8.115) if and only if there exist matrices $X = X^T$ and $Y = Y^T$ satisfying*

$$E^T X E \geq 0, \qquad (8.125)$$

$$E^T Y E \geq 0, \qquad (8.126)$$

$$\begin{bmatrix} A^T X A - E^T X E & A^T X B \\ B^T X A & B^T X B - \gamma I \end{bmatrix} < 0, \qquad (8.127)$$

$$\begin{bmatrix} \mathcal{N} & 0 \\ \hline 0 & I \end{bmatrix} \Gamma \begin{bmatrix} \mathcal{N} & 0 \\ \hline 0 & I \end{bmatrix}^T < 0, \qquad (8.128)$$

$$X - Y \leq 0, \qquad (8.129)$$

and

$$\operatorname{rank}\,(X - Y) \leq \hat{n}, \qquad (8.130)$$

where

$$\Gamma = \left[\begin{array}{cc|c} A^T Y A - E^T Y E & A^T Y B & L^T \\ B^T Y A & B^T Y B - \gamma I & 0 \\ \hline L & 0 & -\gamma I \end{array} \right], \qquad (8.131)$$

$$\mathcal{N} = \begin{bmatrix} C^T \\ D^T \end{bmatrix}^\perp . \qquad (8.132)$$

In this case, the parameters of all desired filters with order $\hat{n}$ corresponding to a feasible solution (X, Y) to (8.125)–(8.130) are given by

$$\begin{bmatrix} \hat{D} & \hat{C} \\ \hat{B} & \hat{A} \end{bmatrix} = \left[-\hat{W}^{-1} \Psi^T \Lambda \Phi_r^T (\Phi_r \Lambda \Phi_r^T)^{-1} \right. $$
$$\left. + \hat{W}^{-1} \Xi^{\frac{1}{2}} \Upsilon (\Phi_r \Lambda \Phi_r^T)^{-\frac{1}{2}} \right] \Phi_l^+ + \Theta - \Theta \Phi_l \Phi_l^+, \qquad (8.133)$$

where

$$\Xi = \hat{W} - \Psi^T \left[\Lambda - \Lambda \Phi_r^T \left(\Phi_r \Lambda \Phi_r^T \right)^{-1} \Phi_r \Lambda \right] \Psi,$$

$$\Lambda = (\Psi \hat{W}^{-1} \Psi^T - \Omega)^{-1},$$

$$\Omega = \begin{bmatrix} A^T X A - E^T Y E & -E^T X_{12} & A^T X B \\ -X_{12}^T E & -X_{22} & 0 \\ B^T X A & 0 & B^T X B - \gamma I \\ L & 0 & 0 \\ 0 & 0 & 0 \\ X_{22}^{-1} X_{12}^T A & 0 & X_{22}^{-1} X_{12}^T B \end{bmatrix}$$

$$\begin{bmatrix} L^T & 0 & A^T X_{12} X_{22}^{-1} \\ 0 & 0 & 0 \\ 0 & 0 & B^T X_{12} X_{22}^{-1} \\ -(\epsilon + \gamma)I & -\epsilon I & 0 \\ -\epsilon I & -\epsilon I & 0 \\ 0 & 0 & -X_{22}^{-1} \end{bmatrix},$$

$$\Psi = \begin{bmatrix} 0\ 0\ 0\ 0\ I\ 0 \\ 0\ 0\ 0\ 0\ 0\ I \end{bmatrix}^T,$$

$$\Phi = \begin{bmatrix} C\ 0\ D\ 0\ 0\ 0 \\ 0\ I\ 0\ 0\ 0\ 0 \end{bmatrix},$$

and Θ is any matrix with appropriate dimensions; Υ is any matrix satisfying $\|\Upsilon\| < 1$; Φ_l and Φ_r are any full rank factors of Φ; that is, $\Phi = \Phi_l \Phi_r$; moreover, $X_{12} \in \mathbb{R}^{n \times \hat{n}}$, $X_{22} \in \mathbb{R}^{\hat{n} \times \hat{n}}$, $X_{22} > 0$ and $\hat{W} > 0$ satisfying $\Lambda > 0$ and

$$Y - X = X_{12} X_{22}^{-1} X_{12}^T \geq 0,$$

and $\epsilon > 0$ is any scalar satisfying

$$\begin{bmatrix} A^T X A - E^T X E & A^T X B & L^T \\ B^T X A & B^T X B - \gamma I & 0 \\ L & 0 & -(\gamma + \epsilon)I \end{bmatrix} < 0.$$

Proof. By Lemma 8.2, it is easy to see that a necessary and sufficient condition for the error dynamic system in (8.119) and (8.120) to satisfy the requirements (RD1) and (RD2) is that there exists a matrix $\tilde{P} = \tilde{P}^T$ such that

$$\tilde{E}^T \tilde{P} \tilde{E} \geq 0, \qquad (8.134)$$

$$\begin{bmatrix} \tilde{A}^T \tilde{P} \tilde{A} - \tilde{E}^T \tilde{P} \tilde{E} & \tilde{A}^T \tilde{P} \tilde{B} & \tilde{C}^T \\ \tilde{B}^T \tilde{P} \tilde{A} & \tilde{B}^T \tilde{P} \tilde{B} - \gamma I & \tilde{D}^T \\ \tilde{C} & \tilde{D} & -\gamma I \end{bmatrix} < 0. \qquad (8.135)$$

Write $\tilde{P}$ as

$$\tilde{P} = \begin{bmatrix} W & W_1 \\ W_1^T & W_2 \end{bmatrix} \qquad (8.136)$$

where the partition is compatible with $\tilde{A}$. Then, (8.134) can be written as

$$\begin{bmatrix} E^T W E & E^T W_1 \\ W_1^T E & W_2 \end{bmatrix} \geq 0, \qquad (8.137)$$

which implies

$$E^T W E \geq 0, \qquad (8.138)$$

$$W_2 \geq 0. \qquad (8.139)$$

Considering (8.135), we have

$$\tilde{A}^T \tilde{P} \tilde{A} - \tilde{E}^T \tilde{P} \tilde{E} < 0. \tag{8.140}$$

It follows from this inequality that

$$\hat{A}^T W_2 \hat{A} - W_2 < 0,$$

which together with (8.139) implies $W_2 > 0$. Then, applying Schur complement to (8.137) gives

$$E^T \left(W - W_1 W_2^{-1} W_1^T \right) E \geq 0. \tag{8.141}$$

Note that

$$\tilde{B} = \bar{B} + \bar{F} G \bar{J}, \quad \tilde{D} = \bar{S} G \bar{J}, \tag{8.142}$$

where

$$\bar{B} = \begin{bmatrix} B \\ 0 \end{bmatrix}, \quad \bar{J} = \begin{bmatrix} D \\ 0 \end{bmatrix}.$$

Then, by substituting $\tilde{A}$, $\tilde{C}$ in (8.71) and $\tilde{B}$, $\tilde{D}$ in (8.142) into (8.135), we have

$$\begin{bmatrix} \left(\bar{A} + \bar{F} G \bar{H}\right)^T \tilde{P} \left(\bar{A} + \bar{F} G \bar{H}\right) - \tilde{E}^T \tilde{P} \tilde{E} & \left(\bar{A} + \bar{F} G \bar{H}\right)^T \tilde{P} \left(\bar{B} + \bar{F} G \bar{J}\right) & \left(\bar{C} + \bar{S} G \bar{H}\right)^T \\ \left(\bar{B} + \bar{F} G \bar{J}\right)^T \tilde{P} \left(\bar{A} + \bar{F} G \bar{H}\right) & \left(\bar{B} + \bar{F} G \bar{J}\right)^T \tilde{P} \left(\bar{B} + \bar{F} G \bar{J}\right) - \gamma I & \left(\bar{S} G \bar{J}\right)^T \\ \bar{C} + \bar{S} G \bar{H} & \bar{S} G \bar{J} & -\gamma I \end{bmatrix} < 0,$$

which can be rewritten as

$$\begin{bmatrix} \bar{A}^T \tilde{P} \bar{A} - \tilde{E}^T \tilde{P} \tilde{E} & \bar{A}^T \tilde{P} \bar{B} & \bar{C}^T \\ \bar{B}^T \tilde{P} \bar{A} & \bar{B}^T \tilde{P} \bar{B} - \gamma I & 0 \\ \bar{C} & 0 & -\gamma I \end{bmatrix} + \begin{bmatrix} \bar{A}^T \tilde{P} \bar{F} \\ \bar{B}^T \tilde{P} \bar{F} \\ \bar{S} \end{bmatrix} \begin{bmatrix} \bar{G} \bar{H} & \bar{G} \bar{J} & 0 \end{bmatrix}$$

$$+ \begin{bmatrix} \bar{H}^T \bar{G}^T \\ \bar{J}^T \bar{G}^T \\ 0 \end{bmatrix} \begin{bmatrix} \bar{F}^T \tilde{P} \bar{A} & \bar{F}^T \tilde{P} \bar{B} & \bar{S}^T \end{bmatrix} + \begin{bmatrix} \bar{H}^T \bar{G}^T \\ \bar{J}^T \bar{G}^T \\ 0 \end{bmatrix} \bar{F}^T \tilde{P} \bar{F} \begin{bmatrix} \bar{G} \bar{H} & \bar{G} \bar{J} & 0 \end{bmatrix}$$

$$< 0. \tag{8.143}$$

Observe that

$$\bar{F}^T \tilde{P} \bar{F} = \begin{bmatrix} 0 & 0 \\ 0 & W_2 \end{bmatrix} \geq 0.$$

Then, it can be seen that (8.143) holds if and only if there exists a scalar $\epsilon > 0$ such that

$$\begin{bmatrix} \bar{A}^T \tilde{P}\bar{A} - \tilde{E}^T \tilde{P}\tilde{E} & \bar{A}^T \tilde{P}\bar{B} & \bar{C}^T \\ \bar{B}^T \tilde{P}\bar{A} & \bar{B}^T \tilde{P}\bar{B} - \gamma I & 0 \\ \bar{C} & 0 & -\gamma I \end{bmatrix} + \begin{bmatrix} \bar{A}^T \tilde{P}\bar{F} \\ \bar{B}^T \tilde{P}\bar{F} \\ \bar{S} \end{bmatrix} \begin{bmatrix} \bar{H}^T \bar{G}^T \\ \bar{J}^T \bar{G}^T \\ 0 \end{bmatrix}^T$$

$$+ \begin{bmatrix} \bar{H}^T \bar{G}^T \\ \bar{J}^T \bar{G}^T \\ 0 \end{bmatrix} \begin{bmatrix} \bar{A}^T \tilde{P}\bar{F} \\ \bar{B}^T \tilde{P}\bar{F} \\ \bar{S} \end{bmatrix}^T + \begin{bmatrix} \bar{H}^T \bar{G}^T \\ \bar{J}^T \bar{G}^T \\ 0 \end{bmatrix} U \begin{bmatrix} \bar{H}^T \bar{G}^T \\ \bar{J}^T \bar{G}^T \\ 0 \end{bmatrix}^T < 0 \quad (8.144)$$

where

$$U = \begin{bmatrix} \epsilon^{-1} I & 0 \\ 0 & W_2 \end{bmatrix}.$$

By some algebraic manipulations, it can be verified that the inequality in (8.144) can be rewritten as

$$\begin{bmatrix} \bar{A}^T \tilde{P}\bar{A} - \tilde{E}^T \tilde{P}\tilde{E} & \bar{A}^T \tilde{P}\bar{B} & \bar{C}^T \\ \bar{B}^T \tilde{P}\bar{A} & \bar{B}^T \tilde{P}\bar{B} - \gamma I & 0 \\ \bar{C} & 0 & -\gamma I \end{bmatrix}$$

$$+ \left(\begin{bmatrix} \bar{H}^T \bar{G}^T \\ \bar{J}^T \bar{G}^T \\ 0 \end{bmatrix} + \begin{bmatrix} \bar{A}^T \tilde{P}\bar{F} \\ \bar{B}^T \tilde{P}\bar{F} \\ \bar{S} \end{bmatrix} U^{-1} \right) U \left(\begin{bmatrix} \bar{H}^T \bar{G}^T \\ \bar{J}^T \bar{G}^T \\ 0 \end{bmatrix} + \begin{bmatrix} \bar{A}^T \tilde{P}\bar{F} \\ \bar{B}^T \tilde{P}\bar{F} \\ \bar{S} \end{bmatrix} U^{-1} \right)^T$$

$$- \begin{bmatrix} \bar{A}^T \tilde{P}\bar{F} \\ \bar{B}^T \tilde{P}\bar{F} \\ \bar{S} \end{bmatrix} U^{-1} \begin{bmatrix} \bar{A}^T \tilde{P}\bar{F} \\ \bar{B}^T \tilde{P}\bar{F} \\ \bar{S} \end{bmatrix}^T < 0, \quad (8.145)$$

which, by Schur complement, can be shown to be equivalent to

$$\begin{bmatrix} \bar{A}^T \bar{P}\bar{A} - \tilde{E}^T \tilde{P}\tilde{E} & \bar{A}^T \bar{P}\bar{B} \\ \bar{B}^T \bar{P}\bar{A} & \bar{B}^T \bar{P}\bar{B} - \gamma I \\ \bar{C} - \bar{S}U^{-1}\bar{F}^T \tilde{P}\bar{A} & -\bar{S}U^{-1}\bar{F}^T \tilde{P}\bar{B} \\ \bar{G}\bar{H} + U^{-1}\bar{F}^T \tilde{P}\bar{A} & \bar{G}\bar{J} + U^{-1}\bar{F}^T \tilde{P}\bar{B} \end{bmatrix}$$

$$\begin{bmatrix} \bar{C}^T - \bar{A}^T \tilde{P}\bar{F}U^{-1}\bar{S}^T & \bar{H}^T \bar{G}^T + \bar{A}^T \tilde{P}\bar{F}U^{-1} \\ -\bar{B}^T \tilde{P}\bar{F}U^{-1}\bar{S}^T & \bar{J}^T \bar{G}^T + \bar{B}^T \tilde{P}\bar{F}U^{-1} \\ -\bar{S}U^{-1}\bar{S}^T - \gamma I & \bar{S}U^{-1} \\ U^{-1}\bar{S}^T & -U^{-1} \end{bmatrix} < 0, \quad (8.146)$$

where

$$\bar{P} = \tilde{P} - \tilde{P}\bar{F}U^{-1}\bar{F}^T \tilde{P} = \begin{bmatrix} W - W_1 W_2^{-1} W_1^T & 0 \\ 0 & 0 \end{bmatrix}.$$

Note that $\bar{F}U^{-1}\bar{S}^T = 0$. Then, we can write (8.146) as

$$\Omega_d + \Psi_d G \Phi_d + (\Psi_d G \Phi_d)^T < 0, \quad (8.147)$$

where

$$
\Omega_d =
\begin{bmatrix}
\bar{A}^T \bar{P} \bar{A} - \tilde{E}^T \tilde{P} \tilde{E} & \bar{A}^T \bar{P} \bar{B} \\
\bar{B}^T \bar{P} \bar{A} & \bar{B}^T \bar{P} \bar{B} - \gamma I \\
\bar{C} & 0 \\
U^{-1} \bar{F}^T \tilde{P} \bar{A} & U^{-1} \bar{F}^T \tilde{P} \bar{B}
\end{bmatrix}
$$

$$
\begin{matrix}
\bar{C}^T & \bar{A}^T \tilde{P} \bar{F} U^{-1} \\
0 & \bar{B}^T \tilde{P} \bar{F} U^{-1} \\
-\bar{S} U^{-1} \bar{S}^T - \gamma I & \bar{S} U^{-1} \\
U^{-1} \bar{S}^T & -U^{-1}
\end{matrix}
\Bigg],
$$

$$
\Psi_d = \begin{bmatrix} 0 & 0 & 0 & I \end{bmatrix}^T,
$$
$$
\Phi_d = \begin{bmatrix} \bar{H} & \bar{J} & 0 & 0 \end{bmatrix}.
$$

By Lemma 7.2, it can be seen that a necessary and sufficient condition for LMI (8.147) to have a solution G is that the following two inequalities hold:

$$
\Psi_d^{\perp} \Omega_d \Psi_d^{\perp T} < 0, \tag{8.148}
$$
$$
\Phi_d^{T\perp} \Omega_d \Phi_d^{T\perp T} < 0. \tag{8.149}
$$

Observe that

$$
\Psi_d^{\perp} =
\begin{bmatrix}
I & 0 & 0 & 0 \\
0 & I & 0 & 0 \\
0 & 0 & I & 0
\end{bmatrix},
$$

$$
\Phi_1^{T\perp} =
\left[
\begin{array}{c|ccc}
\mathcal{N} & 0 & 0 & 0 \\
\hline
0 & 0 & I & 0 \\
0 & 0 & 0 & I
\end{array}
\right]
\left[
\begin{array}{cc|ccc}
I & 0 & 0 & 0 & 0 \\
0 & 0 & I & 0 & 0 \\
0 & I & 0 & 0 & 0 \\
0 & 0 & 0 & I & 0 \\
0 & 0 & 0 & 0 & I
\end{array}
\right].
$$

Therefore, (8.148) and (8.149) give

$$
\begin{bmatrix}
A^T \left(W - W_1 W_2^{-1} W_1^T\right) A - E^T W E & -E^T W_1 \\
-W_1^T E & -W_2 \\
B^T \left(W - W_1 W_2^{-1} W_1^T\right) A & 0 \\
L & 0
\end{bmatrix}
$$

$$
\begin{matrix}
A^T \left(W - W_1 W_2^{-1} W_1^T\right) B & L^T \\
0 & 0 \\
B^T \left(W - W_1 W_2^{-1} W_1^T\right) B - \gamma I & 0 \\
0 & -(\epsilon + \gamma)I
\end{matrix}
\Bigg] < 0, \tag{8.150}
$$

and

$$
\begin{bmatrix} \mathcal{N} & 0 \\ 0 & I \end{bmatrix} \tilde{\Gamma} \begin{bmatrix} \mathcal{N} & 0 \\ 0 & I \end{bmatrix}^T < 0, \tag{8.151}
$$

where

$$\tilde{\Gamma} = \left[\begin{array}{cc|ccc} \tilde{\Gamma}_{11} & \tilde{\Gamma}_{12} & L^T & 0 & A^T W_1 W_2^{-1} \\ \tilde{\Gamma}_{12}^T & \tilde{\Gamma}_{22} & 0 & 0 & B^T W_1 W_2^{-1} \\ \hline L & 0 & -(\epsilon+\gamma)I & -\epsilon I & 0 \\ 0 & 0 & -\epsilon I & -\epsilon I & 0 \\ W_2^{-1} W_1^T A & W_2^{-1} W_1^T B & 0 & 0 & -W_2^{-1} \end{array}\right],$$

$$\tilde{\Gamma}_{11} = A^T \left(W - W_1 W_2^{-1} W_1^T\right) A - E^T \left(W - W_1 W_2^{-1} W_1^T\right) E,$$
$$\tilde{\Gamma}_{12} = A^T \left(W - W_1 W_2^{-1} W_1^T\right) B,$$
$$\tilde{\Gamma}_{22} = B^T \left(W - W_1 W_2^{-1} W_1^T\right) B - \gamma I.$$

Now, set

$$X = W - W_1 W_2^{-1} W_1^T, \quad Y = W.$$

Then, by Schur complement, we have that (8.150) and (8.151) provide the LMIs in (8.127) and (8.128), respectively. Furthermore, it is easy to see that (8.141) and (8.138) give (8.125) and (8.126), respectively, while the inequalities in (8.129) and (8.130) can be obtained by noting

$$X - Y = -W_1 W_2^{-1} W_1^T,$$

and $0 < W_2 \in \mathbb{R}^{\hat{n} \times \hat{n}}$. Finally, when the inequalities (8.125)–(8.130) are feasible, the parametrization of all $\hat{n}$-th order filters corresponding to a feasible solution can be obtained by using the results in [55] and [84]. This completes the proof.   □

Remark 8.8. Theorem 8.4 presents a necessary and sufficient condition for the solvability of the reduced-order H_∞ filtering problem for discrete singular systems. Based on this theorem, it is easy to see that the optimal reduced-order H_∞ filtering problem can be solved by finding a solution to the following optimization problem.

$$\min_{\gamma>0,\ X=X^T, Y=Y^T} \gamma$$

subject to the constraints in (8.125)–(8.130).   ◁

Remark 8.9. In the case when $E = I$, that is, the discrete singular system in (8.113)–(8.115) reduces to a discrete state-space one, we can show that Theorem 8.4 reduces to the Theorem 5 in [66]. Therefore, Theorem 8.4 in this chapter extends existing results on the reduced-order H_∞ filtering problem for discrete state-space systems to the singular case.   ◁

For the case of the zeroth-order H_∞ filtering problem, we have the following simple results, which is free from the rank constraint.

Theorem 8.5. *There exists a filter in the form of (8.124) to solve the zeroth-order H_∞ filtering problem for the discrete singular system in (8.113)–(8.115) if and only if there exists a matrix $X = X^T$ satisfying*

$$E^T X E \geq 0, \qquad (8.152)$$

$$\begin{bmatrix} A^T X A - E^T X E & A^T X B \\ B^T X A & B^T X B - \gamma I \end{bmatrix} < 0, \qquad (8.153)$$

$$\begin{bmatrix} \mathcal{N} & 0 \\ \hline 0 & I \end{bmatrix} \hat{\Gamma} \begin{bmatrix} \mathcal{N} & 0 \\ \hline 0 & I \end{bmatrix}^T < 0, \qquad (8.154)$$

where $\mathcal{N}$ is given in (8.132) and

$$\hat{\Gamma} = \begin{bmatrix} A^T X A - E^T X E & A^T X B & L^T \\ \hline B^T X A & B^T X B - \gamma I & 0 \\ \hline L & 0 & -\gamma I \end{bmatrix}.$$

In this case, all the desired zeroth-order filters in (8.124) corresponding to a feasible solution X to (8.152)–(8.154) are given by

$$\hat{D} = \left[-\hat{W}^{-1} \hat{\Psi}_d^T \Lambda \hat{\Phi}_{dr}^T (\hat{\Phi}_{dr} \Lambda \hat{\Phi}_{dr}^T)^{-1} \right.$$
$$\left. + \hat{W}^{-1} \Xi^{\frac{1}{2}} \Upsilon (\hat{\Phi}_{dr} \Lambda \hat{\Phi}_{dr}^T)^{-\frac{1}{2}} \right] \hat{\Phi}_{dl}^+ + \Theta_d - \Theta_d \hat{\Phi}_{dl} \hat{\Phi}_{dl}^+$$

where

$$\Xi = \hat{W} - \hat{\Psi}_d^T \left[\Lambda - \Lambda \hat{\Phi}_{dr}^T \left(\hat{\Phi}_{dr} \Lambda \hat{\Phi}_{dr}^T \right)^{-1} \hat{\Phi}_{dr} \Lambda \right] \hat{\Psi}_d,$$

$$\Lambda = (\hat{\Psi}_d \hat{W}^{-1} \hat{\Psi}_d^T - \hat{\Omega}_d)^{-1},$$

$$\hat{\Omega}_d = \begin{bmatrix} A^T X A - E^T X E & A^T X B & L^T \\ B^T X A & B^T X B - \gamma I & 0 \\ L & 0 & -\gamma I \end{bmatrix},$$

$$\hat{\Psi}_d = \begin{bmatrix} 0 & 0 & -I \end{bmatrix}^T,$$

$$\hat{\Phi}_d = \begin{bmatrix} C & D & 0 \end{bmatrix},$$

and Θ_d is any matrix with appropriate dimensions; Υ is any matrix satisfying $\|\Upsilon\| < 1$ and $\hat{W} > 0$ such that $\Lambda > 0$; $\hat{\Phi}_{dl}$ and $\hat{\Phi}_{dr}$ are any full rank factors of $\hat{\Phi}_d$; that is, $\hat{\Phi}_d = \hat{\Phi}_{dl} \hat{\Phi}_{dr}$.

Proof. The proof can be carried out by using Lemma 8.2 and following a similar line as in the proof of Theorem 8.4. □

Remark 8.10. The optimal zeroth-order H_∞ filtering problem for the discrete singular system in (8.113)–(8.115) can be solved by finding a solution to the following convex optimization problem.

$$\min_{\gamma > 0,\ X = X^T} \gamma$$

subject to the constraints (8.152)–(8.154).   ◁

8.4 Conclusion

The problem of H_∞ filtering for singular systems has been studied in this chapter. First, in the context of continuous singular systems, the full-order H_∞ filtering problem has been addressed. A linear filter has been designed such that the resulting error system is regular, impulse-free and stable while the closed-loop transfer function satisfies a prescribed H_∞-norm bound constraint. An LMI approach has been developed to solve this problem. It has been shown that the desired filters can be chosen with McMillan degree no more than the number of the exponential modes of the plant. Second, we have investigated the reduced-order H_∞ filtering problem, which is the design of a linear filter with a specified order lower than the order of the system under consideration. Both the continuous and discrete cases have been considered, and necessary and sufficient conditions for the solvability of the reduced-order H_∞ problem have been obtained in terms of certain LMIs and a coupling non-convex rank constraint. For the special case of the zeroth-order H_∞ filtering problem, it has been established that the solvability condition is equivalent to a convex LMI feasibility problem, which then can be easily handled by resorting to some recently developed algorithms for solving convex optimization problems. Part of the results in this chapter have appeared in [187].

9

Delay Systems

9.1 Introduction

Time delays are frequently encountered in many practical engineering systems, such as communication, electronics, hydraulic and chemical systems [90, 134]. It has been shown that time delay is one of the main causes of instability and poor performance of a control system [44, 90, 110]. Therefore, there has been an increasing interest in the control and estimation for time-delay systems; many results for time delay systems have been reported in the literature; see [72, 90, 109, 181], and the references therein.

In this chapter, we address the control and filtering problems for singular systems with time delays. First, the problem of stability analysis is considered. Specifically, for continuous singular delay systems, a sufficient condition guaranteeing a singular delay system to be regular, impulse-free and stable, is proposed in terms of LMIs, while for discrete singular delay systems, a sufficient condition ensuring a singular delay system to be regular, causal and stable is presented. Based on the proposed stability conditions, stabilization results are obtained for continuous and discrete singular delay systems, respectively. Then, for continuous singular delay systems, versions of the bounded real lemma are provided in terms of LMIs, and the H_∞ control problem is solved via an LMI approach. Finally, we investigate the H_∞ filtering problem for continuous singular delay systems. A linear filter is designed, which ensures the resulting error system is admissible and the closed-loop transfer function of the filtering error system satisfies a prescribed H_∞ performance level. An LMI approach is employed to solve this problem and a set of the desired filters is also given in terms of a feasible solution set of the proposed LMI condition.

9.2 Stability

In this section, we are concerned with the stability analysis of singular delay systems. The definitions of regularity, non-impulsiveness (in the continuous case) or causality (in the discrete case) and stability for singular delay systems are proposed. For both continuous and singular delay systems, stability conditions are presented in terms of LMIs.

9.2.1 Continuous Case

Consider a linear continuous singular system with state delay described by

$$E\dot{x}(t) = Ax(t) + A_d x(t - \tau), \tag{9.1}$$

$$x(t) = \phi(t), \quad t \in (-\tau, 0], \tag{9.2}$$

where $x(t) \in \mathbb{R}^n$ is the state of the system. The matrix $E \in \mathbb{R}^{n \times n}$ may be singular; we shall assume that $\text{rank}(E) = r \leq n$. A and A_d are known real constant matrices with appropriate dimensions; the scalar $\tau > 0$ is a constant time delay; $\phi(t)$ is a compatible vector valued continuous function.

It is noted that the singular delay system in (9.1) and (9.2) may have an impulsive solution; however, the regularity and the non-impulsiveness of the pair (E, A) can ensure the existence and uniqueness of an impulse-free solution to the singular delay system in (9.1) and (9.2), which is shown in the following lemma.

Lemma 9.1. *Suppose the pair (E, A) is regular and impulse-free, then the solution to the singular delay system in (9.1) and (9.2) exists and is impulse-free and unique on $(0, \infty)$.*

Proof. Noting the regularity and the non-impulsiveness of the pair (E, A) and using the decomposition as in [36], the desired result follows immediately. $\square$

In view of this, we introduce the following definition for the singular delay system in (9.1) and (9.2).

Definition 9.1.

(I) *The singular delay system in (9.1) and (9.2) is said to be regular and impulse-free if the pair (E, A) is regular and impulse-free.*

(II) The singular delay system in (9.1) and (9.2) is said to be stable if, for any $\varepsilon > 0$, there exists a scalar $\delta(\varepsilon) > 0$ such that, for any compatible initial conditions $\phi(t)$ satisfying

$$\sup_{-\tau < t \leq 0} \|\phi(t)\| < \delta(\varepsilon),$$

the solution $x(t)$ to (9.1) and (9.2) satisfies

$$\|x(t)\| < \varepsilon,$$

and a scalar $\delta > 0$ can be chosen such that

$$\sup_{-\tau < t \leq 0} \|\phi(t)\| < \delta,$$

implies

$$x(t) \to 0, \quad t \to \infty.$$

(III) The singular delay system in (9.1) and (9.2) is said to be admissible if it is regular, impulse-free and stable.

Now, we are in a position to present an admissibility condition for the singular delay system in (9.1) and (9.2).

Theorem 9.1. *The continuous singular delay system in (9.1) and (9.2) is admissible if there exist matrices $Q > 0$ and P such that the following LMIs hold:*

$$E^T P = P^T E \geq 0, \tag{9.3}$$

$$\begin{bmatrix} A^T P + P^T A + Q & P^T A_d \\ A_d^T P & -Q \end{bmatrix} < 0. \tag{9.4}$$

To prove Theorem 9.1, we need the following results.

Lemma 9.2. *Consider the function $\varphi : \mathbb{R}^+ \to \mathbb{R}$. If $\dot{\varphi}$ is bounded on $[0, \infty)$; that is, there exists a scalar $\alpha > 0$ such that $|\dot{\varphi}(t)| \leq \alpha$ for all $t \in [0, \infty)$, then φ is uniformly continuous on $[0, \infty)$.*

Lemma 9.3. *[93] Consider the function $\varphi : \mathbb{R}^+ \to \mathbb{R}$. If φ is uniformly continuous and*

$$\int_0^\infty \varphi(s)ds < \infty,$$

then

$$\lim_{t \to \infty} \varphi(t) = 0.$$

Proof of Theorem 9.1. Under the condition of the theorem, it is easy to see that (9.4) implies

$$A^T P + P^T A < 0. \tag{9.5}$$

Then, by Theorem 2.1, it follows from (9.3) and (9.5) that the pair (E, A) is regular and impulse-free. Therefore, considering Definition 9.1, we have that the singular delay system in (9.1) and (9.2) is regular and impulse-free. Next, we shall show the stability of this singular delay system. To this end, we note that the regularity and the non-impulsiveness of the pair (E, A) imply that there exist two invertible matrices $M \in \mathbb{R}^{n \times n}$ and $N \in \mathbb{R}^{n \times n}$ such that

$$\bar{E} \triangleq MEN = \begin{bmatrix} I_r & 0 \\ 0 & 0 \end{bmatrix}, \quad \bar{A} \triangleq MAN = \begin{bmatrix} A_1 & 0 \\ 0 & I_{n-r} \end{bmatrix}, \tag{9.6}$$

where $I_r \in \mathbb{R}^{r \times r}$ and $I_{n-r} \in \mathbb{R}^{(n-r) \times (n-r)}$ are identity matrices, $A_1 \in \mathbb{R}^{r \times r}$. Now, write

$$\bar{A}_d \triangleq MA_dN = \begin{bmatrix} A_{d11} & A_{d12} \\ A_{d21} & A_{d22} \end{bmatrix}, \tag{9.7}$$

$$\bar{P} \triangleq M^{-T}PN = \begin{bmatrix} \bar{P}_{11} & \bar{P}_{12} \\ \bar{P}_{21} & \bar{P}_{22} \end{bmatrix}, \tag{9.8}$$

$$\bar{Q} \triangleq N^T QN = \begin{bmatrix} \bar{Q}_{11} & \bar{Q}_{12} \\ \bar{Q}_{12}^T & \bar{Q}_{22} \end{bmatrix}, \tag{9.9}$$

where the partitions are compatible with that of $\bar{A}$ in (9.6). Then, from (9.3) and (9.4), and the notations in (9.6)–(9.9), we have

$$\bar{E}^T \bar{P} = \bar{P}^T \bar{E} \geq 0 \tag{9.10}$$

$$\begin{bmatrix} \bar{A}^T \bar{P} + \bar{P}^T \bar{A} + \bar{Q} & \bar{P}^T \bar{A}_d \\ \bar{A}_d^T \bar{P} & -\bar{Q} \end{bmatrix} < 0. \tag{9.11}$$

Noting the expression of $\bar{E}$ in (9.6) and using (9.10), we can deduce that

$$\bar{P}_{11} = \bar{P}_{11}^T \geq 0,$$

and

$$\bar{P}_{12} = 0.$$

Therefore, $\bar{P}$ reduces to

$$\bar{P} = \begin{bmatrix} \bar{P}_{11} & 0 \\ \bar{P}_{21} & \bar{P}_{22} \end{bmatrix}. \tag{9.12}$$

Now, substituting (9.6)–(9.9) and (9.12) into (9.11), we obtain

$$\begin{bmatrix} \bar{P}_{11}A_1 + A_1^T \bar{P}_{11} + \bar{Q}_{11} & \bar{P}_{21}^T + \bar{Q}_{12} \\ \bar{P}_{21} + \bar{Q}_{12}^T & \bar{P}_{22} + \bar{P}_{22}^T + \bar{Q}_{22} \\ A_{d11}^T \bar{P}_{11} + A_{d21}^T \bar{P}_{21} & A_{d21}^T \bar{P}_{22} \\ A_{d12}^T \bar{P}_{11} + A_{d22}^T \bar{P}_{21} & A_{d22}^T \bar{P}_{22} \end{bmatrix}$$

$$\begin{bmatrix} \bar{P}_{11}A_{d11} + \bar{P}_{21}^T A_{d21} & \bar{P}_{11}A_{d12} + \bar{P}_{21}^T A_{d22} \\ \bar{P}_{22}^T A_{d21} & \bar{P}_{22}^T A_{d22} \\ -\bar{Q}_{11} & -\bar{Q}_{12} \\ -\bar{Q}_{12}^T & -\bar{Q}_{22} \end{bmatrix} < 0, \tag{9.13}$$

which implies

$$\begin{bmatrix} \bar{P}_{22} + \bar{P}_{22}^T + \bar{Q}_{22} & \bar{P}_{22}^T A_{d22} \\ A_{d22}^T \bar{P}_{22} & -\bar{Q}_{22} \end{bmatrix} < 0. \tag{9.14}$$

Pre- and post-multiplying (9.14) by

$$\begin{bmatrix} -A_{d22}^T & I \end{bmatrix},$$

and its transpose, respectively, result in

$$A_{d22}^T \bar{Q}_{22} A_{d22} - \bar{Q}_{22} < 0.$$

This together with $\bar{Q}_{22} > 0$ implies

$$\rho(A_{d22}) < 1. \tag{9.15}$$

Now, let

$$\zeta(t) = \begin{bmatrix} \zeta_1(t) \\ \zeta_2(t) \end{bmatrix} = N^{-1}x(t),$$

where

$$\zeta_1(t) \in \mathbb{R}^r, \quad \zeta_2(t) \in \mathbb{R}^{n-r}.$$

Using the expressions in (9.6)–(9.9), the singular delay system in (9.1) can be rewritten as

$$\dot{\zeta}_1(t) = A_1\zeta_1(t) + A_{d11}\zeta_1(t-\tau) + A_{d12}\zeta_2(t-\tau), \tag{9.16}$$
$$0 = \zeta_2(t) + A_{d21}\zeta_1(t-\tau) + A_{d22}\zeta_2(t-\tau). \tag{9.17}$$

Then, it is easy to see that the stability of the singular delay system in (9.1) is equivalent to that of the system in (9.16) and (9.17). In view of this, next we shall prove that the system in (9.16) and (9.17) is stable. To this end, we note that $\bar{P}_{11} = \bar{P}_{11}^T \geq 0$ and (9.13) imply

$$\bar{P}_{11}A_1 + A_1^T \bar{P}_{11} + \bar{Q}_{11} < 0.$$

Therefore, we have $\bar{P}_{11} > 0$. Define

$$V(\zeta_t) = \zeta_1(t)^T \bar{P}_{11}\zeta_1(t) + \int_{t-\tau}^t \zeta(s)^T \bar{Q}\zeta(s)ds,$$

where

$$\zeta_t = \zeta(t+\beta), \quad \beta \in (-\tau, 0].$$

Then, the time-derivative of $V(\zeta_t)$ along the solution of (9.16) and (9.17) is given by

$$
\begin{aligned}
\dot{V}(\zeta_t) &= \frac{d}{dt}\left[\zeta(t)^T \bar{P}^T \bar{E}\zeta(t)\right] + \zeta(t)^T \bar{Q}\zeta(t) - \zeta(t-\tau)^T \bar{Q}\zeta(t-\tau) \\
&= 2\zeta(t)^T \bar{P}^T \bar{A}\zeta(t) + \zeta(t)^T \bar{Q}\zeta(t) \\
&\quad + 2\zeta(t)^T \bar{P}^T \bar{A}_d\zeta(t-\tau) - \zeta(t-\tau)^T \bar{Q}\zeta(t-\tau) \\
&\le \zeta(t)^T \left(\bar{A}^T \bar{P} + \bar{P}^T \bar{A} + \bar{P}^T \bar{A}_d \bar{Q}^{-1} \bar{A}_d^T \bar{P} + \bar{Q}\right)\zeta(t).
\end{aligned}
\tag{9.18}
$$

Applying Schur complement to (9.11) gives

$$
\bar{A}^T \bar{P} + \bar{P}^T \bar{A} + \bar{P}^T \bar{A}_d \bar{Q}^{-1} \bar{A}_d^T \bar{P} + \bar{Q} < 0.
$$

This together with (9.18) implies $\dot{V}(\zeta_t) < 0$ and

$$
\begin{aligned}
&\lambda_1 \left|\zeta_1(t)\right|^2 - V(\zeta_0) \\
&\le \zeta_1(t)^T \bar{P}_{11}\zeta_1(t) - V(\zeta_0) \\
&\le \zeta_1(t)^T \bar{P}_{11}\zeta_1(t) + \int_{t-\tau}^t \zeta(s)^T \bar{Q}\zeta(s)ds - V(\zeta_0) \\
&= \int_0^t \dot{V}(\zeta_s)ds \\
&\le -\lambda_2 \int_0^t \left|\zeta(s)\right|^2 ds \\
&\le -\lambda_2 \int_0^t \left|\zeta_1(s)\right|^2 ds \\
&< 0,
\end{aligned}
\tag{9.19}
$$

where

$$
\begin{aligned}
\lambda_1 &= \lambda_{\min}\left(\bar{P}_{11}\right) > 0, \\
\lambda_2 &= -\lambda_{\max}\left(\bar{A}^T \bar{P} + \bar{P}^T \bar{A} + \bar{P}^T \bar{A}_d \bar{Q}^{-1} \bar{A}_d^T \bar{P} + \bar{Q}\right) > 0.
\end{aligned}
$$

Taking into account (9.19), we can deduce

$$
\lambda_1 \left|\zeta_1(t)\right|^2 + \lambda_2 \int_0^t \left|\zeta_1(s)\right|^2 ds \le V(\zeta_0).
$$

Therefore,

$$
\left|\zeta_1(t)\right|^2 \le m_1,
\tag{9.20}
$$

and

$$
\int_0^t \left|\zeta_1(s)\right|^2 ds \le m_2,
\tag{9.21}
$$

where

$$m_1 = \frac{1}{\lambda_1} V(\zeta_0) > 0, \quad m_2 = \frac{1}{\lambda_2} V(\zeta_0) > 0.$$

Thus, $|\zeta_1(t)|$ is bounded. Considering this and (9.15), it can be deduced from (9.17) that $|\zeta_2(t)|$ is bounded, and hence, it follows from (9.16) that $|\dot{\zeta}_1(t)|$ is bounded, therefore, $\frac{d}{dt} |\zeta_1(t)|^2$ is also bounded. By Lemma 9.2 we have that $|\zeta_1(t)|^2$ is uniformly continuous. Therefore, noting (9.21) and using Lemma 9.3, we obtain

$$\lim_{t \to \infty} |\zeta_1(t)| = 0.$$

It then follows from (9.15) and (9.17) that

$$\lim_{t \to \infty} |\zeta_2(t)| = 0.$$

Thus, the system in (9.16) and (9.17) is stable. Finally, as we have shown that this system is also regular and impulse-free, by Definition 9.1, we then have that the system in (9.16) and (9.17) is admissible. This completes the proof.

$\square$

Remark 9.1. Theorem 9.1 provides a sufficient condition for the singular delay system in (9.1) and (9.2) to be admissible. When $E = I$, the singular delay system in (9.1) and (9.2) reduces to a state-space delay system. In this case, it is easy to show that Theorem 9.1 coincides with Lemma 1 in [103]. Therefore, Theorem 9.1 can be viewed as an extension of existing results on state-space delay systems to singular delay systems. $\triangleleft$

Similar to the derivation of Corollary 2.1, we can also obtain the following admissibility result from Theorem 9.1.

Corollary 9.1. *The singular delay system in (9.1) and (9.2) is admissible if there exist matrices $Q > 0$ and P such that the following LMIs hold:*

$$EP = P^T E^T \geq 0,$$
$$\begin{bmatrix} P^T A^T + AP + Q & A_d P \\ P^T A_d^T & -Q \end{bmatrix} < 0.$$

It is noted that the results in both Theorem 9.1 and Corollary 9.1 involve non-strict LMIs. However, with these results, we can obtain the following admissibility results in terms of strict LMIs.

Theorem 9.2. *The singular delay system in (9.1) and (9.2) is admissible if there exist matrices $P > 0$, $Q > 0$ and Q_1 such that*

$$\begin{bmatrix} A^T \left(PE + SQ_1\right) + \left(PE + SQ_1\right)^T A + Q & \left(PE + SQ_1\right)^T A_d \\ A_d^T \left(PE + SQ_1\right) & -Q \end{bmatrix} < 0,$$

where $S \in \mathbb{R}^{n \times (n-r)}$ is any matrix with full column rank and satisfies $E^T S = 0$.

Corollary 9.2. *The singular delay system in (9.1) and (9.2) is admissible if there exist matrices $P > 0$, $Q > 0$ and Q_1 such that*

$$\begin{bmatrix} \left(PE^T + SQ_1\right)^T A^T + A \left(PE^T + SQ_1\right) + Q & A_d \left(PE^T + SQ_1\right) \\ \left(PE^T + SQ_1\right)^T A_d^T & -Q \end{bmatrix} < 0,$$

(9.22)

where $S \in \mathbb{R}^{n \times (n-r)}$ is any matrix with full column rank and satisfies $ES = 0$.

9.2.2 Discrete Case

Consider a linear discrete singular system with state delay described by

$$Ex(k+1) = Ax(k) + A_d x(k - \tau), \tag{9.23}$$
$$x(i) = \phi(i), \quad i = -\tau, -\tau + 1, \ldots, 0, \tag{9.24}$$

where $x(k) \in \mathbb{R}^n$ is the state. The matrix $E \in \mathbb{R}^{n \times n}$ may be singular; we shall assume that $\operatorname{rank}(E) = r \leq n$. A and A_d are known real constant matrices with appropriate dimensions; the scalar $\tau > 0$ is an integer which represents a constant time delay of the system; $\phi(i)$, $i = -\tau, -\tau + 1, \ldots, 0$, is a compatible vector valued sequence.

For the discrete singular delay system in (9.23) and (9.24), we introduce the following definition.

Definition 9.2.

(I) *The discrete singular delay system in (9.23) and (9.24) is said to be regular if $\det(z^{\tau+1} E - z^\tau A - A_d)$ is not identically zero.*

(II) *The discrete singular delay system in (9.23) and (9.24) is said to be causal if it is regular and $\deg(z^{n\tau} \det(zE - A - z^{-\tau} A_d)) = n\tau + \operatorname{rank}(E)$.*

(III) *The discrete singular delay system in (9.23) and (9.24) is said to be stable if it is regular and $\rho(E, A, A_d) < 1$, where*

$$\rho(E, A, A_d) \triangleq \max_{\lambda \in \{z \mid \det(z^{\tau+1} E - z^\tau A - A_d) = 0\}} |\lambda|.$$

(IV) The discrete singular delay system in (9.23) and (9.24) is said to be admissible if it is regular, causal and stable.

The following theorem presents a sufficient condition for the admissibility of the discrete singular delay system in (9.23).

Theorem 9.3. *The discrete singular delay system in (9.23) is admissible if there exist matrices $Q > 0$ and $P = P^T$ such that*

$$E^T PE \geq 0, \tag{9.25}$$

$$\begin{bmatrix} A^T PA - E^T PE + Q & A^T PA_d \\ A_d^T PA & A_d^T PA_d - Q \end{bmatrix} < 0. \tag{9.26}$$

Proof. Applying Schur complement to (9.26), we obtain

$$Q - A_d^T PA_d > 0, \tag{9.27}$$

and

$$A^T PA - E^T PE + A^T PA_d (Q - A_d^T PA_d)^{-1} A_d^T PA + Q < 0. \tag{9.28}$$

Then, by (9.27) and (9.28), it is easy to show that there exists a scalar $1 > \epsilon > 0$ such that

$$(1 - \epsilon)Q - A_d^T PA_d > 0, \tag{9.29}$$

and

$$A^T PA - E^T PE + A^T PA_d \left[(1 - \epsilon)Q - A_d^T PA_d \right]^{-1} A_d^T PA + Q < 0. \tag{9.30}$$

Define

$$\xi(k) = \left[x(k)^T \ x(k-1)^T \dots x(k-\tau)^T \right]^T.$$

Then, it can be verified that system (9.23) can be rewritten as

$$\hat{E}\xi(k+1) = \hat{A}\xi(k), \tag{9.31}$$

where

$$\hat{E} = \left[\begin{array}{c|c} E & 0_{n \times n\tau} \\ \hline 0_{n\tau \times n} & I_{n\tau \times n\tau} \end{array} \right], \quad \hat{A} = \left[\begin{array}{c|c} A \ 0_{n \times n(\tau-1)} & A_d \\ \hline I_{n\tau \times n\tau} & 0_{n\tau \times n} \end{array} \right]. \tag{9.32}$$

Set

$$\hat{P} = \text{diag}\,(P, Q, Q_1, \dots, Q_{\tau-1}), \tag{9.33}$$

where

$$Q_i = \left(1 - \frac{i\epsilon}{\tau - 1} \right) Q, \quad i = 1, 2, \dots, \tau - 1.$$

It can be shown that

$$
\hat{A}^T \hat{P} \hat{A} - \hat{E}^T \hat{P} \hat{E}
$$

$$
= \begin{bmatrix}
A^T PA - E^T PE + Q & & & -\frac{\epsilon}{\tau-1}Q \\
& & & & -\frac{\epsilon}{\tau-1}Q \\
& & & & \\
A_d^T PA & & & \\
& A^T PA_d & & \\
& & \ddots & \\
-\frac{\epsilon}{\tau-1}Q & & & \\
& A_d^T PA_d - (1-\epsilon)Q
\end{bmatrix}
$$

Using Schur complement, it follows from (9.29), (9.30) and the above equality that

$$
\hat{A}^T \hat{P} \hat{A} - \hat{E}^T \hat{P} \hat{E} < 0. \tag{9.34}
$$

Then, by noting that

$$
\hat{E}^T \hat{P} \hat{E} \geq 0
$$

and (9.34), and employing Theorem 2.3, we have that the discrete singular system in (9.31) is admissible. This implies that there exists a scalar $z \in \mathbb{C}$ such that

$$
\det(z\hat{E} - \hat{A}) \neq 0. \tag{9.35}
$$

If $z = 0$, then it can be deduced from (9.35) that $\det(A_d) \neq 0$, therefore,

$$
\det(z^{\tau+1}E - z^\tau A - A_d) = (-1)^n \det(A_d) \neq 0.
$$

If $z \neq 0$, then, by noting that

$$
\det(z\hat{E} - \hat{A}) = z^{n\tau} \det(zE - A - z^{-\tau}A_d), \tag{9.36}
$$

we have

$$
\det(z^{\tau+1}E - z^\tau A - A_d) \neq 0.
$$

Thus, by Definition 9.2, it follows that the discrete singular delay system in (9.23) is regular. On the other hand, from the causality of the discrete singular system in (9.31) we have

$$
\deg(\det(z\hat{E} - \hat{A})) = \text{rank}(\hat{E}) = n\tau + \text{rank}(E).
$$

This together with (9.36) implies that the discrete singular delay system in (9.23) is causal. Now, from (9.36) and the stability of the discrete singular system (9.31), it can be seen that system (9.23) is stable. Finally, by Definition 9.2, we have that the discrete singular delay system in (9.23) is admissible. This completes the proof.   □

Remark 9.2. In the case when $E = I$, that is, (9.23) reduces to a state-space discrete delay system, it is easy to show that Theorem 9.3 coincides with Theorem 1 in [166]. Therefore, Theorem 9.3 can be viewed as an extension of existing results on stability for state-space discrete delay systems to singular discrete delay systems.   ◁

The following theorem provides a strict LMI condition ensuring the discrete singular delay system in (9.23) to be admissible.

Theorem 9.4. *The discrete singular delay system in (9.23) is admissible if there exist matrices $P > 0$, $Q > 0$ and Q_1 such that*

$$\begin{bmatrix} A^T PA - E^T PE + Q & A^T PA_d \\ A_d^T PA & A_d^T PA_d - Q \end{bmatrix}$$
$$+ \begin{bmatrix} A^T \\ A_d^T \end{bmatrix} SW^T + WS^T \begin{bmatrix} A^T \\ A_d^T \end{bmatrix}^T < 0, \tag{9.37}$$

where $S \in \mathbb{R}^{n \times (n-r)}$ is any matrix with full column rank and satisfies $E^T S = 0$.

Proof. Under the condition of the theorem, we choose

$$\hat{S} = \begin{bmatrix} S \\ 0_{n\tau \times (n-r)} \end{bmatrix}.$$

It is easy to see that $\hat{S}$ is a matrix with full column rank and satisfies $\hat{E}^T \hat{S} = 0$, where $\hat{E}$ is given in (9.32). Now, write

$$W = \begin{bmatrix} W_1 \\ W_2 \end{bmatrix},$$

where $W_1 \in \mathbb{R}^{n \times (n-r)}$ and $W_2 \in \mathbb{R}^{n \times (n-r)}$, and set

$$\hat{W} = \begin{bmatrix} W_1 \\ 0_{n(\tau-1) \times (n-r)} \\ W_2 \end{bmatrix}.$$

Then, it can be seen that

$$\hat{A}^T \hat{P} \hat{A} - \hat{E}^T \hat{P} \hat{E} + \hat{W} \hat{S}^T \hat{A} + \hat{A}^T \hat{S} \hat{W}^T$$
$$= \begin{bmatrix} A^T PA - E^T PE + Q + A^T SW_1^T + W_1 S^T A \\ \\ \\ A_d^T PA + A_d^T SW_1^T + W_2 S^T A \end{bmatrix} \begin{matrix} \\ -\frac{\epsilon}{\tau-1}Q \\ -\frac{\epsilon}{\tau-1}Q \\ \\ \end{matrix}$$

$$A^T P A_d + W_1 S^T A_d + A^T S W_2^T$$

$$\left. \begin{matrix} & \ddots & \\ -\frac{\epsilon}{\tau-1} Q & & \\ & A_d^T P A_d - (1 - \epsilon)Q + A_d^T S W_2^T + W_2 S^T A_d \end{matrix} \right], \qquad (9.38)$$

where $\hat{E}$ and $\hat{A}$ are given in (9.32), and $\hat{P}$ is given in (9.33). Now, considering (9.37) and (9.38), we have

$$\hat{A}^T \hat{P} \hat{A} - \hat{E}^T \hat{P} \hat{E} + \hat{W} \hat{S}^T \hat{A} + \hat{A}^T \hat{S} \hat{W}^T < 0.$$

Therefore, by applying Theorem 2.4, we have that the discrete singular system in (9.31) is admissible. Then, following a similar line as in the proof of Theorem 9.3, we can show that the discrete singular delay system in (9.23) is admissible. □

9.3 State Feedback Control

In this section, we study the stabilization problem for singular delay systems. Attention will be focused on the design of a state feedback controller such that the closed-loop system is admissible. In both the continuous and discrete cases, sufficient conditions for the solvability of the stabilization problem will be obtained.

9.3.1 Continuous Case

Consider a linear continuous singular delay system described by

$$E\dot{x}(t) = Ax(t) + A_d x(t - \tau) + Bu(t), \qquad (9.39)$$
$$x(t) = \phi(t), \quad t \in (-\tau, 0], \qquad (9.40)$$

where $x(t) \in \mathbb{R}^n$ is the state of the system; $u(t) \in \mathbb{R}^m$ is the control input. The matrix $E \in \mathbb{R}^{n \times n}$ may be singular; we shall assume that rank$(E) = r \leq n$. A, A_d and B are known real constant matrices with appropriate dimensions; the scalar $\tau > 0$ is a constant time delay; $\phi(t)$ is a compatible vector valued continuous function.

For the continuous singular delay system in (9.39) and (9.40), we consider the following state feedback controller:

$$u(t) = Kx(t), \quad K \in \mathbb{R}^{m \times n}. \tag{9.41}$$

Applying this controller to (9.39) and (9.40) results in a closed-loop system as follows:

$$E\dot{x}(t) = (A + BK)x(t) + A_d x(t - \tau), \tag{9.42}$$

$$x(t) = \phi(t), \quad t \in (-\tau, 0]. \tag{9.43}$$

Then, we have the following stabilization result.

Theorem 9.5. *Given the continuous singular delay system in (9.39) and (9.40). There exists a state feedback controller (9.41) such that the closed-loop system in (9.42) and (9.43) is admissible if there exist matrices $P > 0$, $Q > 0$, Q_1 and Y such that*

$$\begin{bmatrix} \Pi(P, Q_1) + BY + Y^T B^T + Q & A_d \Omega(P, Q_1) \\ \Omega(P, Q_1)^T A_d^T & -Q \end{bmatrix} < 0, \tag{9.44}$$

where

$$\Pi(P, Q_1) = \Omega(P, Q_1)^T A^T + A\Omega(P, Q_1),$$
$$\Omega(P, Q_1) = PE^T + SQ_1,$$

and $S \in \mathbb{R}^{n \times (n-r)}$ is any matrix with full column rank and satisfies $ES = 0$. In this case, we can assume that the matrix $\Omega(P, Q_1)$ is nonsingular (if this is not the case, then we can choose some $\theta \in (0, 1)$ such that $\hat{\Omega}(P, Q_1) = \Omega(P, Q_1) + \theta\tilde{P}$ is nonsingular and satisfies (9.44), in which $\tilde{P}$ is any nonsingular matrix satisfying $E\tilde{P} = \tilde{P}^T E^T \geq 0$), then a desired stabilizing state feedback controller can be chosen as

$$u(t) = Y\Omega(P, Q_1)^{-1} x(t). \tag{9.45}$$

Proof. Let

$$A_c = A + BY\Omega(P, Q_1)^{-1}.$$

Then, it is easy to see that (9.44) can be rewritten as

$$\begin{bmatrix} A_c \Omega(P, Q_1) + \Omega(P, Q_1)^T A_c^T + Q & A_d \Omega(P, Q_1) \\ \Omega(P, Q_1)^T A_d^T & -Q \end{bmatrix} < 0.$$

Therefore, by Corollary 9.2, it is easy to have that the closed-loop system (9.42) with the controller in (9.45) is admissible. $\qquad \square$

9.3.2 Discrete Case

Consider a linear discrete singular system with state delay described by

$$Ex(k+1) = Ax(k) + A_d x(k-\tau) + Bu(k), \tag{9.46}$$
$$x(i) = \phi(i), \quad i = -\tau, -\tau+1, \dots, 0. \tag{9.47}$$

For the discrete singular delay system in (9.46) and (9.47), we consider the following state feedback controller:

$$u(k) = Kx(k), \quad K \in \mathbb{R}^{m \times n}. \tag{9.48}$$

Applying this to (9.46) results in the closed-loop system as follows:

$$Ex(k+1) = (A + BK)x(k) + A_d x(k-\tau). \tag{9.49}$$

Then, we have the following stabilization result.

Theorem 9.6. *Given the discrete singular delay system in (9.46) and (9.47). There exists a state feedback controller in (9.48) such that the closed-loop system in (9.49) is admissible if there exist a scalar $\delta > 0$, matrices $P > 0$, $Q > 0$, W_1 and W_2 such that*

$$\Omega = Q - A_d^T P A_d - A_d^T S W_2^T - W_2 S^T A_d > 0, \tag{9.50}$$

and

$$\Gamma_1 - \Gamma_2 \Gamma_3^{-1} \Gamma_2^T < 0, \tag{9.51}$$

where $S \in \mathbb{R}^{n \times (n-r)}$ is any matrix with full column rank and satisfies $E^T S = 0$, and

$$\Gamma_1 = A^T P A + W_1 S^T A + A^T S W_1^T - E^T P E + Q$$
$$+ \left(A^T J^T + W_1 S^T A_d\right) \Omega^{-1} \left(A^T J^T + W_1 S^T A_d\right)^T,$$
$$\Gamma_2 = \left[\left(A^T J^T + W_1 S^T A_d\right) \Omega^{-1} J + A^T P + W_1 S^T\right] B,$$
$$\Gamma_3 = B^T \left(J^T \Omega^{-1} J + P\right) B + \delta I,$$
$$J = \left(A_d^T P + W_2 S^T\right) A.$$

In this case, a desired stabilizing state feedback control law can be chosen as

$$u(k) = -\Gamma_3^{-1} \Gamma_2^T x(k). \tag{9.52}$$

Proof. Let

$$A_c = A + BK.$$

Then, it can be verified that

$$
\begin{aligned}
& A_c^T P A_c - E^T P E + Q + A_c^T S W_1^T + W_1 S^T A_c \\
& + \left(A_c^T J^T + W_1 S^T A_d \right) \Omega^{-1} \left(A_c^T J^T + W_1 S^T A_d \right)^T \\
& = \Gamma_1 + K^T B^T \left(J^T \Omega^{-1} J + P \right) B K + \Gamma_2 K + K^T \Gamma_2^T \\
& \le \Gamma_1 + K^T \Gamma_3 K + \Gamma_2 K + K^T \Gamma_2^T .
\end{aligned}
\tag{9.53}
$$

Now, under the controller in (9.52), we substitute

$$
K = -\Gamma_3^{-1} \Gamma_2^T ,
$$

into (9.53) to have

$$
\begin{aligned}
& A_c^T P A_c - E^T P E + Q + A_c^T S W_1^T + W_1 S^T A_c \\
& + \left(A_c^T J^T + W_1 S^T A_d \right) \Omega^{-1} \left(A_c^T J^T + W_1 S^T A_d \right)^T \\
& \le \Gamma_1 - \Gamma_2 \Gamma_3^{-1} \Gamma_2^T \\
& < 0,
\end{aligned}
$$

which, by Schur complement, implies

$$
\begin{bmatrix}
\Lambda & A_c^T J^T + W_1 S^T A_d \\
J A + A_d^T S W_1^T & -\Omega
\end{bmatrix}
< 0,
\tag{9.54}
$$

where
$$
\Lambda = A_c^T P A_c - E^T P E + Q + A_c^T S W_1^T + W_1 S^T A_c.
$$

Noting (9.54) and using Theorem 9.4, we have that under the controller in (9.52), the closed-loop system in (9.49) is admissible. This completes the proof. $\square$

9.4 H_∞ Control of Continuous Delay Systems

In this section, we are concerned with the problem of H_∞ control for continuous singular delay systems. The class of singular delay systems to be considered is described by

$$
\begin{aligned}
E\dot{x}(t) &= Ax(t) + A_d x(t - \tau) + Bu(t) + D\omega(t), & (9.55) \\
z(t) &= Cx(t) + C_d x(t - \tau) + Hu(t), & (9.56) \\
x(t) &= \phi(t), \quad t \in (-\tau, 0], & (9.57)
\end{aligned}
$$

where $x(t) \in \mathbb{R}^n$ is the state; $u(t) \in \mathbb{R}^m$ is the control input; $\omega(t) \in \mathbb{R}^p$ is the disturbance input which belongs to $\mathcal{L}_2[0, \infty)$; and $z(t) \in \mathbb{R}^s$ is the controlled output. The matrix $E \in \mathbb{R}^{n \times n}$ may be singular; we shall assume that $\operatorname{rank}(E) = r \leq n$. The matrices A, A_d, B, C, C_d, D and H are known real constant matrices with appropriate dimensions. The scalar $\tau > 0$ is a constant time delay; $\phi(t)$ is a compatible vector valued continuous function.

Now consider the following linear state feedback controller

$$u(t) = Kx(t), \quad K \in \mathbb{R}^{m \times n}. \tag{9.58}$$

Then the closed-loop system from (9.55), (9.56) and (9.58) can be written as

$$E\dot{x}(t) = A_c x(t) + A_d x(t - \tau) + Dw(t), \tag{9.59}$$
$$z(t) = C_c x(t) + C_d x(t - \tau), \tag{9.60}$$

where

$$A_c = A + BK, \quad C_c = C + HK. \tag{9.61}$$

When the pair (E, A_c) is regular, the closed-loop transfer function matrix $G_{zw}(s)$ from the disturbance $w(t)$ to the controlled output $z(t)$ is well defined and given by

$$G_{zw}(s) = \left(C_c + C_{dc} e^{-s\tau}\right) \left(sE - A_c - A_{dc} e^{-s\tau}\right)^{-1} D. \tag{9.62}$$

Then, the H_∞ control problem to be addressed can be formulated as follows: given a continuous singular delay system in (9.55)–(9.57) and a scalar $\gamma > 0$, design a state feedback controller in the form of (9.58) such that the closed-loop system in (9.59) and (9.60) is admissible when $w(t) = 0$ while the H_∞ norm of the closed-loop transfer function $G_{zw}(s)$ satisfies

$$\|G_{zw}\|_\infty < \gamma. \tag{9.63}$$

Before presenting a condition for the solvability of the H_∞ control problem, we first provide a version of the bounded real lemma for continuous singular delay systems, which will play an important role in solving the H_∞ control problem. For this purpose, we consider the unforced singular delay system of (9.55) and (9.56); that is,

$$E\dot{x}(t) = Ax(t) + A_d x(t - \tau) + Dw(t), \tag{9.64}$$
$$z(t) = Cx(t) + C_d x(t - \tau). \tag{9.65}$$

A version of the bounded real lemma for this singular delay system is given in the following lemma.

Lemma 9.4. *Given a scalar $\gamma > 0$. The continuous singular delay system in (9.64) and (9.65) is admissible and its transfer function*

$$G(s) = \left(C + C_d e^{-s\tau}\right)\left(sE - A - A_d e^{-s\tau}\right)^{-1} D, \qquad (9.66)$$

satisfies

$$\|G\|_\infty < \gamma, \qquad (9.67)$$

if there exist matrices $Q > 0$ and P such that

$$E^T P = P^T E \geq 0, \qquad (9.68)$$

$$\begin{bmatrix} A^T P + P^T A + Q & P^T A_d & C^T & P^T D \\ A_d^T P & -Q & C_d^T & 0 \\ C & C_d & -I & 0 \\ D^T P & 0 & 0 & -\gamma^2 I \end{bmatrix} < 0. \qquad (9.69)$$

Proof. From (9.69), it is easy to see

$$\begin{bmatrix} A^T P + P^T A + Q & P^T A_d \\ A_d^T P & -Q \end{bmatrix} < 0.$$

Noting this and (9.68), and then using Theorem 9.1, we have that the singular delay system in (9.64) with $w(t) \equiv 0$ is admissible. Next we shall show the H_∞ performance of the singular delay system in (9.64) and (9.65). To this end, we apply Schur complement to (9.69) and obtain

$$W = Q - C_d^T C_d > 0, \qquad (9.70)$$

and

$$A^T P + P^T A + C^T C + \gamma^{-2} P^T D D^T P + Q$$
$$+ \left(P^T A_d + C^T C_d\right) W^{-1} \left(P^T A_d + C^T C_d\right)^T < 0. \qquad (9.71)$$

By (9.71), it can be shown that there exists a matrix $V > 0$ such that

$$A^T P + P^T A + C^T C + \gamma^{-2} P^T D D^T P + Q$$
$$+ \left(P^T A_d + C^T C_d\right) W^{-1} \left(P^T A_d + C^T C_d\right)^T + V < 0. \qquad (9.72)$$

Noting (9.70), we can obtain

$$\left(P^T A_d + C^T C_d\right) e^{-j\omega\tau} + \left(P^T A_d + C^T C_d\right)^T e^{j\omega\tau}$$
$$\leq \left(P^T A_d + C^T C_d\right) W^{-1} \left(P^T A_d + C^T C_d\right)^T + W.$$

This together with (9.72) implies

$$\left(A^T + A_d^T e^{j\omega\tau}\right) P + P^T \left(A + A_d e^{-j\omega\tau}\right)$$
$$+ \Phi(-j\omega)^T \Phi(j\omega) + S$$
$$\leq A^T P + P^T A + C^T C + S + Q$$
$$+ \left(P^T A_d + C^T C_d\right) W^{-1} \left(P^T A_d + C^T C_d\right)^T$$
$$< 0, \qquad (9.73)$$

where

$$\Phi(j\omega) = C + C_d e^{-j\omega\tau}, \quad S = \gamma^{-2}P^T DD^T P + V.$$

Therefore, from (9.68) and (9.73), we have

$$\Psi(-j\omega)^T P + P^T\Psi(j\omega) - \Phi(-j\omega)^T\Phi(j\omega) - S > 0, \tag{9.74}$$

where

$$\Psi(j\omega) = j\omega E - A - A_d e^{-j\omega\tau}.$$

It then follows from (9.74) that $\Psi(j\omega)$ is invertible for all $\omega \in \mathbb{R}$. That is, $\Psi(j\omega)^{-1}$ is well defined. Now, pre- and post-multiplying (9.74) by $D^T\Psi(-j\omega)^{-T}$ and $\Psi(j\omega)^{-1}D$, respectively, we have

$$D^T P\Psi(j\omega)^{-1}D + D^T\Psi(-j\omega)^{-T}P^T D$$
$$-D^T\Psi(-j\omega)^{-T}\left[\Phi(-j\omega)^T\Phi(j\omega) + S\right]\Psi(j\omega)^{-1}D \geq 0. \tag{9.75}$$

That is,

$$D^T\Psi(-j\omega)^T\Phi(-j\omega)^T\Phi(j\omega)\Psi(j\omega)^{-1}D$$
$$\leq D^T P\Psi(j\omega)^{-1}D + D^T\Psi(-j\omega)^{-T}P^T D - D^T\Psi(-j\omega)^{-T}S\Psi(j\omega)^{-1}D.$$

Using this inequality, we can obtain

$$\gamma^2 I - G(-j\omega)^T G(j\omega)$$
$$= \gamma^2 I - D^T\Psi(-j\omega)^T\Phi(-j\omega)^T\Phi(j\omega)\Psi(j\omega)^{-1}D$$
$$\geq \gamma^2 I - D^T P\Psi(j\omega)^{-1}D - D^T\Psi(-j\omega)^{-T}P^T D$$
$$+ D^T\Psi(-j\omega)^{-T}S\Psi(j\omega)^{-1}D$$
$$= \gamma^2 I - D^T PS^{-1}P^T D$$
$$+ \left[D^T\Psi(-j\omega)^{-T} - D^T PS^{-1}\right]$$
$$\times S\left[D^T\Psi(-j\omega)^{-T} - D^T PS^{-1}\right]^T$$
$$\geq \gamma^2 I - D^T PS^{-1}P^T D. \tag{9.76}$$

On the other hand, we note that

$$\begin{bmatrix} \gamma^2 I & D^T P \\ P^T D & \gamma^{-2}P^T DD^T P + V \end{bmatrix} > 0,$$

which, by Schur complement, implies

$$\gamma^2 I - D^T PS^{-1}P^T D > 0.$$

With this inequality and (9.76), we have that for all $\omega \in \mathbb{R}$,

$$\gamma^2 I - G(-j\omega)^T G(j\omega) > 0.$$

That is, (9.67) is satisfied. This completes the proof. $\square$

Remark 9.3. In the case when $E = I$, that is, the continuous singular delay system in (9.64) and (9.65) reduces to a state-space delay system, it can be shown that Lemma 9.4 coincides with Theorem 1 in [103]. Thus, Lemma 9.4 can be viewed as an extension of existing results on H_∞ performance analysis for state-space delay systems to singular delay systems.   ◁

The bounded real lemma in Lemma 9.4 involves a non-strict LMI (9.68). However, by Lemma 9.4, it is easy to have another version of the bounded real lemma for singular delay systems in terms of a strict LMI, which is provided in the following corollary.

Corollary 9.3. *Given a scalar $\gamma > 0$ and a continuous singular delay system in (9.64) and (9.65). This system is admissible and the transfer function $G(s)$ in (9.66) satisfies (9.67) if there exist matrices $P > 0$, $Q > 0$ and Q_1 such that*

$$\begin{bmatrix} \Theta\left(P,Q_1\right) + Q + \gamma^{-2}DD^T & A_d\Omega\left(P,Q_1\right) & \Omega\left(P,Q_1\right)^T C^T \\ \Omega\left(P,Q_1\right)^T A_d^T & -Q & \Omega\left(P,Q_1\right)^T C_d^T \\ C\Omega\left(P,Q_1\right) & C_d\Omega\left(P,Q_1\right) & -I \end{bmatrix} < 0, \qquad (9.77)$$

where $S \in \mathbb{R}^{n\times(n-r)}$ is any matrix with full column rank and satisfies $ES = 0$, and

$$\Theta\left(P,Q_1\right) = \Omega\left(P,Q_1\right)^T A^T + A\Omega\left(P,Q_1\right), \qquad (9.78)$$
$$\Omega\left(P,Q_1\right) = \left(PE^T + SQ_1\right). \qquad (9.79)$$

Now, by Corollary 9.3, it is easy to have the following H_∞ control result.

Theorem 9.7. *Given a scalar $\gamma > 0$ and a continuous singular delay system in (9.55)–(9.57). There exists a state feedback controller (9.58) such that the closed-loop system in (9.59) and (9.60) is admissible and its transfer function (9.62) satisfies (9.63) if there exist matrices $P > 0$, $Q > 0$, Q_1 and Y such that*

$$\begin{bmatrix} W\left(P,Q_1\right) + Q + \gamma^{-2}DD^T & A_d\Omega\left(P,Q_1\right) & \Omega\left(P,Q_1\right)^T C^T + Y^T H^T \\ \Omega\left(P,Q_1\right)^T A_d^T & -Q & \Omega\left(P,Q_1\right)^T C_d^T \\ C\Omega\left(P,Q_1\right) + HY & C_d\Omega\left(P,Q_1\right) & -I \end{bmatrix} < 0,$$
$$(9.80)$$

where

$$W\left(P,Q_1\right) = \Omega\left(P,Q_1\right)^T A^T + A\Omega\left(P,Q_1\right) + B_1Y + Y^T B_1^T,$$

and $\Omega\left(P,Q_1\right)$ is given in (9.79), $S \in \mathbb{R}^{n\times(n-r)}$ is any matrix with full column rank and satisfies $ES = 0$. In this case, we can assume that the matrix

$\Omega\left(P, Q_1\right)$ *is nonsingular (if this is not the case, then we can choose some* $\theta \in (0, 1)$ *such that* $\hat{\Omega}\left(P, Q_1\right) = \Omega\left(P, Q_1\right) + \theta\tilde{P}$ *is nonsingular and satisfies* *(9.80), in which* $\tilde{P}$ *is any nonsingular matrix satisfying* $E\tilde{P} = \tilde{P}^T E^T \geq 0$*),* *then a desired state feedback controller can be chosen as*

$$u(t) = Y\Omega\left(P, Q_1\right)^{-1} x(t).$$

9.5 H_∞ Filtering of Continuous Delay Systems

In this section, we study the H_∞ filtering problem for continuous singular delay systems. The class of linear singular delay systems to be considered is described by

$$E\dot{x}(t) = Ax(t) + A_d x(t - \tau) + D\omega(t), \tag{9.81}$$
$$y(t) = Cx(t) + C_d x(t - \tau) + G\omega(t), \tag{9.82}$$
$$z(t) = Lx(t) + L_d x(t - \tau), \tag{9.83}$$
$$x(t) = \phi(t), \quad t \in (-\tau, 0], \tag{9.84}$$

where $x(t) \in \mathbb{R}^n$ is the state; $y(t) \in \mathbb{R}^m$ is the measurement; $z(t) \in \mathbb{R}^q$ is the signal to be estimated; $\omega(t) \in \mathbb{R}^p$ is the disturbance input which belongs to $\mathcal{L}_2[0, \infty)$. The matrix $E \in \mathbb{R}^{n \times n}$ may be singular; we shall assume that rank$(E) = r \leq n$. A, A_d, C, C_d, D, G, L, and L_d are known real constant matrices with appropriate dimensions. The scalar $\tau > 0$ is a constant time delay; $\phi(t)$ is a compatible vector valued continuous function.

Now, we consider the following filter for the estimate of $z(t)$:

$$E_f \dot{\hat{x}}(t) = A_f \hat{x}(t) + B_f y(t), \tag{9.85}$$
$$\hat{z}(t) = C_f \hat{x}(t), \tag{9.86}$$

where $\hat{x}(t) \in \mathbb{R}^{\hat{n}}$ and $\hat{z}(t) \in \mathbb{R}^q$ are the state and the output of the filter, respectively. The matrices $E_f \in \mathbb{R}^{\hat{n} \times \hat{n}}$, $A_f \in \mathbb{R}^{\hat{n} \times \hat{n}}$, $B_f \in \mathbb{R}^{\hat{n} \times p}$ and $C_f \in \mathbb{R}^{q \times \hat{n}}$ are to be determined. Let

$$e(t) = \left[x(t)^T \; \hat{x}(t)^T\right]^T, \quad \tilde{z}(t) = z(t) - \hat{z}(t). \tag{9.87}$$

Then, the filtering error dynamics from the system in (9.81)–(9.84) and the filter in (9.85) and (9.86) can be written as:

$$E_c \dot{e}(t) = A_c e(t) + A_{dc} H e(t - \tau) + D_c \omega(t), \tag{9.88}$$
$$\tilde{z}(t) = C_c e(t) + C_{dc} H e(t - \tau), \tag{9.89}$$

where

$$E_c = \begin{bmatrix} E & 0 \\ 0 & E_f \end{bmatrix}, \quad A_c = \begin{bmatrix} A & 0 \\ B_f C & A_f \end{bmatrix}, \tag{9.90}$$

$$A_{dc} = \begin{bmatrix} A_d \\ B_f C_d \end{bmatrix}, \quad D_c = \begin{bmatrix} D \\ B_f G \end{bmatrix}, \tag{9.91}$$

$$C_c = \begin{bmatrix} L & -C_f \end{bmatrix}, \quad C_{dc} = L_d, \tag{9.92}$$

$$H = \begin{bmatrix} I & 0 \end{bmatrix}. \tag{9.93}$$

The purpose of the H_∞ filter problem is the design of a filter in the form of (9.85) and (9.86) such that the filtering error system in (9.88) and (9.89) is admissible and the transfer function from $w(t)$ to $\tilde{z}(t)$ given as

$$G_c(s) = \left(C_c + C_{dc} H e^{-s\tau} \right) \left(sE - A_c - A_{dc} H e^{-s\tau} \right)^{-1} D_c, \tag{9.94}$$

satisfies

$$\|G_c\|_\infty < \gamma, \tag{9.95}$$

where $\gamma > 0$ is a prescribed scalar.

We present a sufficient condition for the solvability of the H_∞ filtering problem in the following theorem.

Theorem 9.8. *Given a continuous singular delay system in (9.81)–(9.84) and a scalar $\gamma > 0$. There exists a filter in the form of (9.85) and (9.86) such that the H_∞ filtering problem is solvable if there exist matrices $Q > 0$, X, Y, Φ, Ψ and Λ such that*

$$E^T X = X^T E \geq 0, \tag{9.96}$$

$$E^T Y = Y^T E \geq 0, \tag{9.97}$$

$$E^T (X - Y) \geq 0,$$

and

$$\begin{bmatrix} A^T Y + Y^T A + Q & \Phi^T + Q \\ \Phi + Q & \Xi + Q \\ A_d^T X & A_d^T X + C_d^T \Psi^T \\ L - \Lambda & L \\ D^T Y & D^T X + G^T \Psi^T \end{bmatrix}$$

$$\begin{matrix} Y^T A_d & L^T - \Lambda^T & Y^T D \\ X^T A_d + \Psi C_d & L^T & X^T D + \Psi G \\ -Q & L_d^T & 0 \\ L_d & -I & 0 \\ 0 & 0 & -\gamma^2 I \end{matrix} \end{bmatrix} < 0, \tag{9.98}$$

where

$$\Xi = X^T A + A^T X + \Psi C + C^T \Psi^T.$$

In this case, there exist nonsingular matrices S, $\tilde{S}$, W and $\tilde{W}$ such that

$$E^T \tilde{S} = S^T E, \tag{9.99}$$
$$EW = \tilde{W}^T E^T, \tag{9.100}$$
$$XY^{-1} = I - \tilde{S}W, \tag{9.101}$$
$$Y^{-1}X = I - \tilde{W}S. \tag{9.102}$$

Then a desired H_∞ filter in the form of (9.85) and (9.86) can be chosen with parameters as

$$E_f = E, \tag{9.103}$$
$$A_f = S^{-T}\left(\Phi - A^T Y - X^T A - \Psi C\right)Y^{-1}W^{-1}, \tag{9.104}$$
$$B_f = S^{-T}\Psi, \tag{9.105}$$
$$C_f = \Lambda Y^{-1}W^{-1}. \tag{9.106}$$

Proof. Similar to the proof of Theorem 8.1, it can be shown that the LMIs in (9.96)–(9.98) guarantee the existence of the nonsingular matrices S, $\tilde{S}$, W and $\tilde{W}$ satisfying (9.99)–(9.102). Now, define

$$\Pi_1 = \begin{bmatrix} \bar{Y} & I \\ W & 0 \end{bmatrix}, \quad \Pi_2 = \begin{bmatrix} I & X \\ 0 & S \end{bmatrix},$$

where $\bar{Y} = Y^{-1}$. It is noted that both Π_1 and Π_2 are nonsingular. Set

$$\hat{P} = \Pi_2 \Pi_1^{-1}.$$

Then, similar to the proof of Theorem 8.1, we can show that $\hat{P}$ is nonsingular, and

$$\hat{E}^T \hat{P} = \hat{P}^T \hat{E} \geq 0, \tag{9.107}$$

where

$$\hat{E} = \mathrm{diag}(E, E).$$

Now, pre- and post-multiplying (9.98) by $\mathrm{diag}(\bar{Y}^T, I, I, I, I)$ and its transpose, respectively, we obtain

$$\left[\begin{array}{cccc} \Pi_1^T A_c^T \hat{P} \Pi_1 + \Pi_1^T \hat{P}^T A_c \Pi_1 + \Pi_1^T H^T Q H \Pi_1 & \Pi_1^T \hat{P}^T A_{dc} & \Pi_1^T C_c^T & \Pi_1^T \hat{P}^T D_c \\ A_{dc}^T \hat{P} \Pi_1 & -Q & C_{dc}^T & 0 \\ C_c \Pi_1 & C_{dc} & -I & 0 \\ D_c^T \hat{P} \Pi_1 & 0 & 0 & -\gamma^2 I \end{array}\right] < 0, \tag{9.108}$$

where the matrices A_c, A_{dc}, B_c, C_c and D_c are given in (9.90)–(9.92) with the parameters A_f, B_f and C_f given in (9.104)–(9.106), respectively. Then, pre- and post-multiplying (9.108) by $\mathrm{diag}(\Pi_1^{-T}, I, I, I)$ and its transpose, respectively, result in

$$\begin{bmatrix} A_c^T \hat{P} + \hat{P}^T A_c + H^T Q H & \hat{P}^T A_{dc} & C_c^T & \hat{P}^T D_c \\ A_{dc}^T \hat{P} & -Q & C_{dc}^T & 0 \\ C_c & C_{dc} & -I & 0 \\ D_c^T \hat{P} & 0 & 0 & -\gamma^2 I \end{bmatrix} < 0.$$

This inequality holds if and only if there exist matrices Q_{12} and $Q_{22} = Q_{22}^T$ with appropriate dimensions such that

$$\hat{Q} = \begin{bmatrix} Q & Q_{12} \\ Q_{12}^T & Q_{22} \end{bmatrix} > 0,$$

and

$$\begin{bmatrix} A_c^T \hat{P} + \hat{P}^T A_c + \hat{Q} & \hat{P}^T A_{dc} H & C_c^T & \hat{P}^T D_c \\ H^T A_{dc}^T \hat{P} & -\hat{Q} & C_{dc}^T & 0 \\ C_c & C_{dc} & -I & 0 \\ D_c^T \hat{P} & 0 & 0 & -\gamma^2 I \end{bmatrix} < 0,$$

which, by Schur complement, implies

$$\begin{bmatrix} A_c^T \hat{P} + \hat{P}^T A_c + \hat{Q} + \gamma^{-2} \hat{P}^T D_c D_c^T \hat{P} & \hat{P}^T A_{dc} H & C_c^T \\ H^T A_{dc}^T \hat{P} & -\hat{Q} & C_{dc}^T \\ C_c & C_{dc} & -I \end{bmatrix} < 0.$$

Noting this, (9.107), and then using Lemma 9.4, we have that the error system in (9.88) and (9.89) resulting from the singular delay system in (9.81)–(9.84) and the filter in the form of (9.85) and (9.86) with the parameters given in (9.103)–(9.106) is admissible and its transfer function satisfies (9.95). This completes the proof.   □

9.6 Conclusion

This chapter has studied the control and filtering problems for singular delay systems. First, sufficient stability conditions for both continuous and discrete singular delay systems have been proposed in terms of LMIs, which guarantee a singular system to be regular, impulse-free (in the continuous case) or causal (in the discrete case) and stable. Based on the proposed stability conditions, the stabilization problem for both continuous and discrete singular delay systems has been solved, respectively. Then, for continuous singular delay systems, versions of the bounded real lemma have been presented, which can be regarded as extensions of the bounded real lemma for state-space delay systems to singular delay systems. Based on these results, the H_∞ control problem has been addressed and state feedback controllers have been designed such that the closed-loop system is admissible and satisfies a prescribed H_∞ performance level. Finally, we have considered the H_∞ filtering problem for continuous singular delay systems. A sufficient condition for the solvability of this problem has been derived in terms of LMIs. When these LMIs are feasible, a set of the parameters of a desired filter has also been given. Part of the results in this chapter have appeared in [188].

10

Markovian Jump Systems

10.1 Introduction

Markovian jump systems can model stochastic systems with abrupt structural variation resulting from the occurrence of some inner discrete events in the system such as failures and repairs of machine in manufacturing systems, modifications of the operating point of a linearized model of a nonlinear system [123]. The study of Markovian jump systems is of both practical and theoretical importance. Therefore, during the past ten years, much attention has been addressed to investigate such a class of stochastic systems, and certain filtering and control issues related to Markovian jump systems have been studied [39, 42, 149]. However, most of the results reported in the literature were obtained in the context of state-space models, which may limit the scope of their applications.

In this chapter, we study Markovian jump singular systems. First, the concept of stochastic admissibility is defined as a Markovian jump singular system is regular, impulse-free (in the continuous case) or causal (in the discrete case), and stochastically stable. Necessary and sufficient conditions for stochastic admissibility are obtained in terms of LMIs for continuous and discrete Markovian jump singular systems, respectively. Then, the problem of guaranteed cost control for continuous Markovian jump singular systems with norm-bounded parameter uncertainties is addressed. State feedback controllers are designed such that the closed-loop system is stochastically admissible and a quadratic cost function is guaranteed to have an upper bound. Finally, we deal with the H_∞ control problem for continuous Markovian jump singular systems. The purpose is to design state feedback controllers which ensure that the closed-loop system is stochastically admissible and satisfies a prescribed H_∞ performance level. A sufficient condition for the solvability of this problem is provided and desired state feedback controllers can be constructed by solving a set of LMIs.

10.2 Stability

In this section, we consider the problem of stability analysis of Markovian jump singular systems. We present conditions under which a Markovian jump singular system is guaranteed to be regular, impulse-free (continuous case) or causal (discrete case) and stochastically stable. All these conditions are given in terms of LMIs.

10.2.1 Continuous Case

Consider a class of continuous singular systems with Markovian jump parameters described by

$$E\dot{x}(t) = A(r_t)x(t), \tag{10.1}$$

where $x(t) \in \mathbb{R}^n$ is the state of the system. The matrix $E \in \mathbb{R}^{n \times n}$ may be singular; we shall assume that $\text{rank}(E) = r \leq n$; $\{r_t\}$ is a continuous-time Markovian process with right continuous trajectories taking values in a finite set

$$\mathcal{S} = \{1, 2, \ldots, \mathcal{N}\}, \tag{10.2}$$

with transition probability matrix $\Pi \triangleq \{\pi_{ij}\}$ given by

$$\Pr\{r_{t+h} = j \,|\, r_t = i\} = \begin{cases} \pi_{ij}h + o(h) & i \neq j, \\ 1 + \pi_{ii}h + o(h) & i = j, \end{cases} \tag{10.3}$$

where

$$h > 0, \quad \lim_{h \to 0} \frac{o(h)}{h} = 0, \tag{10.4}$$

and $\pi_{ij} \geq 0$, for $j \neq i$, is the transition rate from mode i at time t to mode j at time $t + h$, which satisfies

$$\pi_{ii} = - \sum_{j=1,\ j \neq i}^{\mathcal{N}} \pi_{ij}. \tag{10.5}$$

For notational simplicity, the matrix $A(r_t)$ will be denoted by A_i for each possible $r_t = i$, $i \in \mathcal{S}$.

Now, we introduce the following definition for the continuous Markovian jump singular system in (10.1).

Definition 10.1.

(I) *The continuous Markovian jump singular system in (10.1) is said to be regular if $\det(sE - A_i)$ is not identically zero for every $i \in \mathcal{S}$.*

(II) *The continuous Markovian jump singular system in (10.1) is said to be impulse-free if $\deg(\det(sE - A_i)) = \operatorname{rank}(E)$ for every $i \in \mathcal{S}$.*

(III) *The continuous Markovian jump singular system in (10.1) is said to be stochastically stable if for any $x_0 \in \mathbb{R}^n$ and $r_0 \in \mathcal{S}$, there exists a scalar $\tilde{M}(x_0, r_0) > 0$ such that*

$$\lim_{t \to \infty} \mathcal{E} \left\{ \left. \int_0^t x^T(s, x_0, r_0) x(s, x_0, r_0) ds \right| x_0, r_0 \right\} \leq \tilde{M}(x_0, r_0),$$

where $x(t, x_0, r_0)$ denotes the solution to system (10.1) at time t under the initial conditions x_0 and r_0.

(IV) *The continuous Markovian jump singular system in (10.1) is said to be stochastically admissible if it is regular, impulse-free and stochastically stable.*

The following theorem provides a necessary and sufficient condition for the stochastic admissibility of the continuous Markovian jump singular system in (10.1).

Theorem 10.1. *The continuous Markovian jump singular system in (10.1) is stochastically admissible if and only if there exist matrices P_i, $i = 1, 2, \ldots, \mathcal{N}$, such that*

$$E^T P_i = P_i^T E \geq 0, \tag{10.6}$$

$$\sum_{j=1}^{\mathcal{N}} \pi_{ij} E^T P_j + P_i^T A_i + A_i^T P_i < 0. \tag{10.7}$$

To prove this theorem, we need the following result.

Lemma 10.1. [50] *Consider a continuous Markovian jump state-space system described by*

$$\dot{x}(t) = A(r_t)x(t),$$

where the continuous-time Markovian process $\{r_t\}$ takes values in a finite set $\mathcal{S}$ and satisfies (10.3)–(10.5). The system is stochastically stable if and only if there exist matrices $P_i > 0$, $i = 1, 2, \ldots, \mathcal{N}$, such that

$$\sum_{j=1}^{\mathcal{N}} \pi_{ij} P_j + P_i A_i + A_i^T P_i < 0.$$

Proof of Theorem 10.1. (*Sufficiency*) Suppose that there exist matrices P_i, $i = 1, 2, \ldots, \mathcal{N}$, such that (10.6) and (10.7) hold. Under this condition, we first show the regularity and the non-impulsiveness of the Markovian jump singular system in (10.1). To this end, we choose two nonsingular matrices $\hat{M}$ and $\hat{N}$ such that

$$\hat{M} E \hat{N} = \begin{bmatrix} I & 0 \\ 0 & 0 \end{bmatrix}, \tag{10.8}$$

and write

$$\hat{M} A_i \hat{N} = \begin{bmatrix} \hat{A}_{i1} & \hat{A}_{i2} \\ \hat{A}_{i3} & \hat{A}_{i4} \end{bmatrix}, \tag{10.9}$$

$$\hat{M}^{-T} P_i \hat{N} = \begin{bmatrix} \hat{P}_{i1} & \hat{P}_{i2} \\ \hat{P}_{i3} & \hat{P}_{i4} \end{bmatrix}, \tag{10.10}$$

for $i = 1, 2, \ldots, \mathcal{N}$, where the partitions of $\hat{M} A_i \hat{N}$ and $\hat{M}^{-T} P_i N$ are compatible with that of $\hat{M} E \hat{N}$ in (10.8). By (10.6), it can be shown that $\hat{P}_{i2} = 0$ for $i = 1, 2, \ldots, \mathcal{N}$. Pre- and post-multiplying (10.7) by $\hat{N}^T$ and $\hat{N}$, respectively, we have

$$\begin{bmatrix} * & * \\ * & \hat{A}_{i4}^T \hat{P}_{i4} + \hat{P}_{i4}^T \hat{A}_{i4} \end{bmatrix} < 0, \tag{10.11}$$

where $*$ represents a matrix which will not be used in the following discussion. Then, by (10.11), we have that for $i = 1, 2, \ldots, \mathcal{N}$,

$$\hat{A}_{i4}^T \hat{P}_{i4} + \hat{P}_{i4}^T \hat{A}_{i4} < 0, \tag{10.12}$$

which implies $\hat{A}_{i4}$ is nonsingular for $i = 1, 2, \ldots, \mathcal{N}$. Therefore, by (10.12) and Definition 10.1, it is easy to show that the Markovian jump singular system in (10.1) is regular and impulse-free.

Next, we will show the stochastic stability of system (10.1). To this end, we note that $\hat{A}_{i4}$ is nonsingular for any $i \in \mathcal{S}$. Then, we can set

$$\tilde{M}_i = \begin{bmatrix} I & -\hat{A}_{i2} \hat{A}_{i4}^{-1} \\ 0 & I \end{bmatrix} M. \tag{10.13}$$

By (10.8)–(10.10), it is easy to see that

$$\hat{E} = \tilde{M}_i E \hat{N} = \begin{bmatrix} I & 0 \\ 0 & 0 \end{bmatrix}, \tag{10.14}$$

$$\hat{A}_i = \tilde{M}_i A_i \hat{N} = \begin{bmatrix} \tilde{A}_{i1} & 0 \\ \hat{A}_{i3} & \hat{A}_{i4} \end{bmatrix}, \tag{10.15}$$

$$\tilde{M}_i^{-T} P_i \hat{N} = \begin{bmatrix} \hat{P}_{i1} & 0 \\ \tilde{P}_{i3} & \hat{P}_{i4} \end{bmatrix}, \tag{10.16}$$

where
$$\tilde{A}_{i1} = \hat{A}_{i1} - \hat{A}_{i2}\hat{A}_{i4}^{-1}\hat{A}_{i3}, \quad \tilde{P}_{i3} = \hat{P}_{i3} + \hat{A}_{i4}^{-T}\hat{A}_{i2}^{T}\hat{P}_{i1}.$$

Then, for any $i \in \mathcal{S}$, the Markovian jump singular system in (10.1) is restricted system equivalent to

$$\dot{\xi}_1(t) = \tilde{A}_{i1}\xi_1(t), \tag{10.17}$$
$$0 = \hat{A}_{i3}\xi_1(t) + \hat{A}_{i4}\xi_2(t), \tag{10.18}$$

where

$$\xi(t) = \begin{bmatrix} \xi_1(t) \\ \xi_2(t) \end{bmatrix} = \hat{N}^{-1}x(t). \tag{10.19}$$

It follows from (10.6) and (10.7) that $\hat{P}_{i1} > 0$. Now, define

$$V(\xi_1(t), r_t) = \xi_1(t)^T \hat{P}_1(r_t)\xi_1(t).$$

Let $\mathcal{A}$ be the weak infinitesimal generator of the random process $\{\xi_1(t), r_t\}$. Then, for each $r_t = i$, $i \in \mathcal{S}$, we have

$$\mathcal{A}V(\xi_1(t), i) = \xi_1(t)^T \left(\sum_{j=1}^{\mathcal{N}} \pi_{ij}\hat{P}_{j1} + \hat{P}_{i1}\tilde{A}_{i1} + \tilde{A}_{i1}^{T}\hat{P}_{i1} \right) \xi_1(t). \tag{10.20}$$

Now, pre- and post-multiplying (10.7) by $\hat{N}^T$ and $\hat{N}$, respectively, we obtain

$$\sum_{j=1}^{\mathcal{N}} \pi_{ij}\left(\tilde{M}_j E\hat{N}\right)^T \left(\tilde{M}_j^{-T}P_j\hat{N}\right) + \left(\tilde{M}_i^{-T}P_i\hat{N}\right)^T \left(\tilde{M}_i A_i\hat{N}\right)$$
$$+ \left(\tilde{M}_i A_i\hat{N}\right)^T \left(\tilde{M}_i^{-T}P_i\hat{N}\right) < 0. \tag{10.21}$$

Substituting the expressions in (10.14)–(10.16) to (10.21) results in

$$\sum_{j=1}^{\mathcal{N}} \pi_{ij}\hat{P}_{j1} + \hat{P}_{i1}\tilde{A}_{i1} + \tilde{A}_{i1}^{T}\hat{P}_{i1} < 0,$$

which implies that there exists a scalar $a > 0$ such that for $i = 1, 2, \ldots, \mathcal{N}$,

$$\sum_{j=1}^{\mathcal{N}} \pi_{ij}\hat{P}_{j1} + \hat{P}_{i1}\tilde{A}_{i1} + \tilde{A}_{i1}^{T}\hat{P}_{i1} + aI < 0.$$

This together with (10.20) gives

$$\mathcal{A}V(\xi_1(t), i) \le -a\xi_1(t)^T \xi_1(t), \tag{10.22}$$

for $i = 1, 2, \ldots, \mathcal{N}$. Now applying Dynkin's formula to (10.22), we have that for each $r_t = i$, $i \in \mathcal{S}$, $t > 0$,

$$\mathcal{E}\left\{V(\xi_1(t),r_t)|\,\xi_1(0),r_0\right\} - V(\xi_1(0),r_0)$$

$$= \mathcal{E}\left\{\left.\int_0^t \mathcal{A}V(\xi_1(\tau),r_\tau)d\tau\right|\xi_1(0),r_0\right\}$$

$$\leq -a\mathcal{E}\left\{\left.\int_0^t \xi_1(\tau)^T\xi_1(\tau)\,d\tau\right|\xi_1(0),r_0\right\},$$

and this implies

$$\mathcal{E}\left\{\left.\int_0^t \xi_1(\tau)^T\xi_1(\tau)\,d\tau\right|\xi_1(0),r_0\right\} \leq \frac{1}{a}V(\xi_1(0),r_0).$$

With this and (10.18), it is easy to show that there exists a scalar $b(\xi_0,r_0) > 0$ such that

$$\mathcal{E}\left\{\left.\int_0^t \xi(\tau)^T\xi(\tau)\,d\tau\right|\xi(0),r_0\right\} \leq b(\xi_0,r_0).$$

Noting this and (10.19), we can show that there exists a scalar $\tilde{M}(x_0,r_0) > 0$ such that

$$\mathcal{E}\left\{\left.\int_0^t x^T(s,x_0,r_0)x(s,x_0,r_0)ds\right|x_0,r_0\right\} \leq \tilde{M}(x_0,r_0).$$

Therefore, by Definition 10.1, it can be seen that the Markovian jump singular system in (10.1) is stochastically stable. Since it has already been shown that system (10.1) is regular and impulse-free, by Definition 10.1, we have that the Markovian jump singular system in (10.1) is stochastically admissible.

(*Necessity*) Suppose that the continuous Markovian jump singular system in (10.1) is stochastically admissible. Then, we choose two nonsingular matrices $\hat{M}$ and $\hat{N}$ such that (10.8) holds. Now, write $\hat{M}A_i\hat{N}$ in the form of (10.9). Since the continuous Markovian jump singular system in (10.1) is regular and impulse-free, we have that $\hat{A}_{i4}$ is nonsingular for $i = 1,2,\ldots,\mathcal{N}$. Let $\tilde{M}_i$ be given in (10.13). Then, $\hat{E} = \tilde{M}_iE\hat{N}$ and $\hat{A}_i = \tilde{M}_iA_i\hat{N}$ can be written as in (10.14) and (10.15), respectively. It is easy to see that the stochastic stability of (10.1) implies that the state-space Markovian jump system in (10.17) is stochastically stable. Therefore, applying Lemma 10.1 to the state-space Markovian jump system in (10.17) gives that there exist matrices $\bar{P}_i > 0$, $i = 1,2,\ldots,\mathcal{N}$, such that

$$\sum_{j=1}^{\mathcal{N}} \pi_{ij}\bar{P}_j + \bar{P}_i\tilde{A}_{i1} + \tilde{A}_{i1}^T\bar{P}_i < 0. \tag{10.23}$$

Then, it is easy to see that we can always find a sufficiently small scalar $\delta > 0$ such that for $i = 1,2,\ldots,\mathcal{N}$,

$$\sum_{j=1}^{\mathcal{N}} \pi_{ij} \hat{E}^T \begin{bmatrix} \bar{P}_j & 0 \\ 0 & \delta \hat{A}_{j4}^{-T} \end{bmatrix} + \begin{bmatrix} \bar{P}_i & 0 \\ 0 & -\delta \hat{A}_{i4}^{-1} \end{bmatrix} \hat{A}_i + \hat{A}_i^T \begin{bmatrix} \bar{P}_i & 0 \\ 0 & -\delta \hat{A}_{i4}^{-T} \end{bmatrix}$$

$$= \begin{bmatrix} \sum_{j=1}^{\mathcal{N}} \pi_{ij} \bar{P}_j + \bar{P}_i \tilde{A}_{i1} + \tilde{A}_{i1}^T \bar{P}_i & -\delta \hat{A}_{i3}^T \hat{A}_{i4}^{-T} \\ -\delta \hat{A}_{i4}^{-1} \hat{A}_{i3} & -2\delta I \end{bmatrix}$$

$$< 0. \tag{10.24}$$

Pre- and post-multiplying (10.24) by $\hat{N}^{-T}$ and $\hat{N}^{-1}$, respectively, we obtain

$$\sum_{j=1}^{\mathcal{N}} \pi_{ij} \left(\tilde{M}_j^{-1} \hat{E} \hat{N}^{-1} \right)^T \left(\tilde{M}_j^T \begin{bmatrix} \bar{P}_j & 0 \\ 0 & -\delta \hat{A}_{j4}^{-T} \end{bmatrix} \hat{N}^{-1} \right)$$

$$+ \left(\tilde{M}_i^T \begin{bmatrix} \bar{P}_i & 0 \\ 0 & -\delta \hat{A}_{i4}^{-T} \end{bmatrix} \hat{N}^{-1} \right)^T P_i \tilde{M}_i \left(\tilde{M}_i^{-1} \hat{A}_i \hat{N}^{-1} \right)$$

$$+ \left(\tilde{M}_i^{-1} \hat{A}_i \hat{N}^{-1} \right)^T \left(\tilde{M}_i^T \begin{bmatrix} \bar{P}_i & 0 \\ 0 & -\delta \hat{A}_{i4}^{-T} \end{bmatrix} \hat{N}^{-1} \right) < 0. \tag{10.25}$$

Now, let

$$P_i = \tilde{M}_i^T \begin{bmatrix} \bar{P}_i & 0 \\ 0 & -\delta \hat{A}_{i4}^{-T} \end{bmatrix} \hat{N}^{-1}. \tag{10.26}$$

Then, by (10.14), (10.15) and (10.26), it is easy to see that a set of matrices P_i, $i = 1, 2, \ldots, \mathcal{N}$, given in (10.26) satisfies (10.6) and (10.7). This completes the proof. $\qquad\square$

Remark 10.1. In the case when $E = I$, the continuous Markovian jump singular system in (10.1) reduces to a state-space Markovian jump system. In this case, Theorem 10.1 coincides with Lemma 10.1. Therefore, Theorem 10.1 can be viewed as an extension of existing results on stochastic stability for state-space Markovian jump systems to singular Markovian jump systems. $\triangleleft$

10.2.2 Discrete Case

Consider a class of discrete singular systems with Markovian jump parameters described by

$$Ex(k+1) = A(r_k)x(k), \tag{10.27}$$

where $x(k) \in \mathbb{R}^n$ is the system state. The matrix $E \in \mathbb{R}^{n \times n}$ may be singular; we shall assume that $\text{rank}(E) = r \leq n$; the parameter r_k represents a discrete-time, discrete-state Markovian chain taking values in a finite set in (10.2) with transition probabilities

$$\Pr \{r_{k+1} = j \,|\, r_k = i\} = \pi_{ij}, \tag{10.28}$$

where $\pi_{ij} \geq 0$, and for any $i \in \mathcal{S}$,

$$\sum_{j=1}^{\mathcal{N}} \pi_{ij} = 1. \tag{10.29}$$

For notational simplicity, in the sequel, for each possible $r_k = i$, $i \in \mathcal{S}$, the matrix $A(r_k)$ will be denoted by A_i.

Now, we introduce the following definition for the discrete Markovian jump singular system in (10.27).

Definition 10.2.

(I) *The discrete Markovian jump singular system in (10.27) is said to be regular if, for each $i \in \mathcal{S}$, $\det(zE - A_i)$ is not identically zero.*

(II) *The discrete Markovian jump singular system in (10.27) is said to be causal if, for each $i \in \mathcal{S}$, $\deg(\det(zE - A_i)) = \operatorname{rank}(E)$.*

(III) *The discrete Markovian jump singular system in (10.27) is said to be stochastically stable if for any $x_0 \in \mathbb{R}^n$ and $r_0 \in \mathcal{S}$, there exists a scalar $\tilde{M}(x_0, r_0) > 0$ such that*

$$\lim_{N \to \infty} \mathcal{E}\left\{ \sum_{k=0}^{N} |x(k, x_0, r_0)|^2 \,\bigg|\, x_0, r_0 \right\} \leq \tilde{M}(x_0, r_0),$$

where $x(k, x_0, r_0)$ denotes the solution to system (10.27) at time k under the initial conditions x_0 and r_0.

(IV) *The discrete Markovian jump singular system in (10.27) is said to be stochastically admissible if it is regular, causal and stochastically stable.*

The following theorem presents a necessary and sufficient condition under which the discrete Markovian jump singular system in (10.27) is guaranteed to be stochastically admissible.

Theorem 10.2. *The discrete Markovian jump singular system in (10.27) is stochastically admissible if and only if there exist matrices $P_i = P_i^T$, $i = 1, 2, \ldots, \mathcal{N}$, such that*

$$E^T P_i E \geq 0, \tag{10.30}$$

$$A_i^T \hat{P}_i A_i - E^T P_i E < 0, \tag{10.31}$$

where

$$\hat{P}_i = \sum_{j=1}^{\mathcal{N}} \pi_{ij} P_j.$$

To prove this theorem, we need the following lemma.

Lemma 10.2. [85] *Consider a discrete Markovian jump state-space system described by*

$$x(k+1) = A(r_k)x(k),$$

where the discrete-time, discrete-state Markovian chain r_k takes values in a finite set S and satisfies (10.28) and (10.29). The system is stochastically stable if and only if there exist matrices $Q_i > 0$, $i = 1, 2, \ldots, \mathcal{N}$, such that

$$A_i^T \hat{Q}_i A_i - Q_i < 0,$$

where

$$\hat{Q}_i = \sum_{j=1}^{\mathcal{N}} \pi_{ij} Q_j.$$

Proof of Theorem 10.2. (*Sufficiency*) Suppose there exist matrices $P_i = P_i^T$, $i = 1, 2, \ldots, \mathcal{N}$, such that (10.30) and (10.31) hold. Under this condition, we first show the regularity and the causality of the Markovian jump singular system in (10.27). To this end, we choose two nonsingular matrices M and N such that

$$E = M \begin{bmatrix} I & 0 \\ 0 & 0 \end{bmatrix} N. \tag{10.32}$$

Write

$$\tilde{P}_i = M^T P_i M = \begin{bmatrix} P_{i1} & P_{i2} \\ P_{i2}^T & P_{i3} \end{bmatrix}, \tag{10.33}$$

$$A_i = M \begin{bmatrix} A_{1i} & A_{2i} \\ A_{3i} & A_{4i} \end{bmatrix} N, \tag{10.34}$$

where the partitions of $\tilde{P}_i$ and A_i are compatible with that of E in (10.32). Pre- and post-multiplying (10.31) by N^{-T} and N^{-1}, respectively, and then using the notations in (10.33) and (10.34), we obtain

$$\begin{bmatrix} H_{1i} & H_{2i} \\ H_{2i}^T & H_{3i} \end{bmatrix} < 0, \tag{10.35}$$

where

$$H_{1i} = A_{1i}^T \left(\sum_{j=1}^{\mathcal{N}} \pi_{ij} P_{j1} \right) A_{1i} + A_{3i}^T \left(\sum_{j=1}^{\mathcal{N}} \pi_{ij} P_{j2}^T \right) A_{1i}$$

$$+ A_{1i}^T \left(\sum_{j=1}^{\mathcal{N}} \pi_{ij} P_{j2} \right) A_{3i} + A_{3i}^T \left(\sum_{j=1}^{\mathcal{N}} \pi_{ij} P_{j3} \right) A_{3i} - P_{i1},$$

$$H_{2i} = A_{1i}^T \left(\sum_{j=1}^{\mathcal{N}} \pi_{ij} P_{j1} \right) A_{2i} + A_{3i}^T \left(\sum_{j=1}^{\mathcal{N}} \pi_{ij} P_{j2}^T \right) A_{2i}$$

$$+ A_{1i}^T \left(\sum_{j=1}^{\mathcal{N}} \pi_{ij} P_{j2} \right) A_{4i} + A_{3i}^T \left(\sum_{j=1}^{\mathcal{N}} \pi_{ij} P_{j3} \right) A_{4i},$$

$$H_{3i} = A_{2i}^T \left(\sum_{j=1}^{\mathcal{N}} \pi_{ij} P_{j1} \right) A_{2i} + A_{2i}^T \left(\sum_{j=1}^{\mathcal{N}} \pi_{ij} P_{j2} \right) A_{4i}$$

$$+ A_{4i}^T \left(\sum_{j=1}^{\mathcal{N}} \pi_{ij} P_{j2}^T \right) A_{2i} + A_{4i}^T \left(\sum_{j=1}^{\mathcal{N}} \pi_{ij} P_{j3} \right) A_{4i}.$$

It is easy to see that (10.35) implies

$$H_{3i} < 0. \tag{10.36}$$

Note that (10.30) gives $P_{j1} \geq 0$ for each $j \in \mathcal{S}$. Then, it follows from (10.36) that the matrix A_{4i} is nonsingular for each $i \in \mathcal{S}$. Thus, by Definition 10.2, it is easy to show that the discrete Markovian jump singular system in (10.27) is regular and causal. Next, we will show the stochastic stability of system (10.27). To this end, we pre- and post-multiply (10.35) by

$$L_i = \begin{bmatrix} I & -A_{3i}^T A_{4i}^{-T} \\ 0 & I \end{bmatrix} \tag{10.37}$$

and its transpose, respectively, and then we can deduce that $P_{i1} > 0$ for $i \in \mathcal{S}$. Now, set

$$\xi(k) = N^{-1} x(k).$$

Then it is easy to see that the stochastic stability of system (10.27) is equivalent to that of the following system

$$\begin{bmatrix} \xi_1(k+1) \\ 0 \end{bmatrix} = \tilde{A}(r_k) \begin{bmatrix} \xi_1(k) \\ \xi_2(k) \end{bmatrix}, \tag{10.38}$$

where

$$\tilde{A}(r_k) = M^{-1} A(r_k) N^{-1},$$

$$\xi(k) = \begin{bmatrix} \xi_1(k)^T & \xi_2(k)^T \end{bmatrix}^T,$$

and $\xi_1(k) \in \mathbb{R}^r$, $\xi_2(k) \in \mathbb{R}^{n-r}$. Now, define

$$V\left(\xi_1(k), r_k\right) = \xi_1(k)^T P_1(r_k)\xi_1(k)$$

$$= \xi(k)^T \begin{bmatrix} I & 0 \\ 0 & 0 \end{bmatrix} \tilde{P}(r_k) \begin{bmatrix} I & 0 \\ 0 & 0 \end{bmatrix} \xi(k),$$

where $\tilde{P}(r_k)$ is given in (10.33). Then, it can be verified that

$$\mathcal{E}\left\{V(\xi_1(k+1), r_{k+1})| \xi_k, r_k = i\right\} - V(\xi_1(k), r_k = i)$$

$$= \xi(k)^T \left\{\tilde{A}_i^T \left(\sum_{j=1}^N \pi_{ij}\tilde{P}_j\right) \tilde{A}_i - \begin{bmatrix} I & 0 \\ 0 & 0 \end{bmatrix} \tilde{P}_i \begin{bmatrix} I & 0 \\ 0 & 0 \end{bmatrix}\right\} \xi(k). \quad (10.39)$$

On the other hand, it follows from (10.31) that

$$\tilde{A}_i^T \left(\sum_{j=1}^N \pi_{ij}\tilde{P}_j\right) \tilde{A}_i - \begin{bmatrix} I & 0 \\ 0 & 0 \end{bmatrix} \tilde{P}_i \begin{bmatrix} I & 0 \\ 0 & 0 \end{bmatrix} < 0,$$

which implies that there exists a scalar $a > 0$ such that

$$\tilde{A}_i^T \left(\sum_{j=1}^N \pi_{ij}\tilde{P}_j\right) \tilde{A}_i - \begin{bmatrix} I & 0 \\ 0 & 0 \end{bmatrix} \tilde{P}_i \begin{bmatrix} I & 0 \\ 0 & 0 \end{bmatrix} + aI < 0.$$

With this and (10.39), we obtain

$$\mathcal{E}\left\{V(\xi_1(k+1), r_{k+1})| \xi_k, r_k = i\right\} - V(\xi_1(k), r_k = i) < -a\left|\xi_1(k)\right|^2.$$

Therefore,

$$\frac{\mathcal{E}\left\{V(\xi_1(k+1), r_{k+1})| \xi_k, r_k\right\} - V(\xi_1(k), r_k)}{V(\xi_1(k), r_k)}$$

$$< -\min_{r_k \in \mathcal{S}} \left(\frac{a}{\lambda_{\max}\left(P_1(r_k)\right)}\right) = \alpha - 1,$$

where

$$0 < \alpha = 1 - \min_{r_k \in \mathcal{S}} \left(\frac{a}{\lambda_{\max}\left(P_1(r_k)\right)}\right) < 1.$$

Then, following the same lines as in the proof of Theorem 2.1 in [85], we can deduce that there exists a scalar $\delta_1 > 0$ such that

$$\lim_{N \to \infty} \mathcal{E}\left\{\sum_{k=0}^N \left|\xi_1(k)\right|^2 \middle| \xi_0, r_0\right\} \leq \delta_1 V\left(\xi_1(0), r_0\right). \quad (10.40)$$

Now, by (10.38), it can be seen that

$$\xi_2(k) = -A_4(r_k)^{-1}A_3(r_k)\xi_1(k).$$

Let

$$\delta_2 = \max_{k \in \mathcal{S}} \left| A_4(r_k)^{-1} A_3(r_k) \right|.$$

Then

$$|\xi_2(k)| \le \delta_2 \, |\xi_1(k)|,$$

and hence

$$\lim_{N \to \infty} \mathcal{E} \left\{ \sum_{k=0}^{N} |\xi(k)|^2 \bigg| \xi_0, r_0 \right\} \le \delta_1 \left(1 + \delta_2^2\right) V\left(\xi_1(0), r_0\right),$$

which, by Definition 10.2, gives that the discrete Markovian jump singular system in (10.27) is stochastically stable. This together with the regularity and causality of the system (10.27) implies that (10.27) is stochastically admissible.

(*Necessity*) Suppose that the discrete Markovian jump singular system in (10.27) is stochastically admissible. Then, we choose two nonsingular matrices M and N such that (10.32) holds. Now, write A_i in the form of (10.34). The regularity and causality of (10.27) imply A_{4i} is nonsingular for any $i \in \mathcal{S}$. Set

$$\tilde{N}_i = N^{-1} L_i^T$$

where L_i is given in (10.37). Then, it can be verified that

$$\hat{E} = M^{-1}E\tilde{N}_i = \begin{bmatrix} I & 0 \\ 0 & 0 \end{bmatrix}, \tag{10.41}$$

$$\hat{A}_i = M^{-1}A_i\tilde{N}_i = \begin{bmatrix} \hat{A}_{1i} & A_{2i} \\ 0 & A_{4i} \end{bmatrix} \tag{10.42}$$

where

$$\hat{A}_{1i} = A_{1i} - A_{2i}A_{4i}^{-1}A_{3i}.$$

It can be seen that the stochastic stability of the discrete Markovian jump singular system in (10.27) implies that the discrete Markovian jump state-space system

$$\varsigma(k+1) = \hat{A}_1(r_t)\varsigma(k),$$

is stochastically stable. Therefore, by Lemma 10.2, we have that there exists matrices $\bar{P}_i > 0$, $i = 1, 2, \ldots, \mathcal{N}$, such that

$$\hat{A}_{1i}^T \check{P}_i \hat{A}_{1i} - \bar{P}_i < 0, \tag{10.43}$$

where

$$\check{P}_i = \sum_{j=1}^{\mathcal{N}} \pi_{ij} \bar{P}_j.$$

It is easy to see that we can always find a sufficiently large scalar $\delta > 0$ such that for $i = 1, 2, \ldots, \mathcal{N}$,

$$\hat{A}_i^T \begin{bmatrix} \check{P}_i & 0 \\ 0 & -\delta I \end{bmatrix} \hat{A}_i - \hat{E}^T \begin{bmatrix} \bar{P}_i & 0 \\ 0 & -\delta I \end{bmatrix} \hat{E}$$

$$= \begin{bmatrix} \hat{A}_{1i}^T \check{P}_i \hat{A}_{1i} - \bar{P}_i & \hat{A}_{1i}^T \check{P}_i A_{2i} \\ A_{2i}^T \check{P}_i \hat{A}_{1i} & -\delta A_{4i}^T A_{4i} \end{bmatrix}$$

$$< 0. \tag{10.44}$$

Pre- and post-multiplying (10.44) by $\tilde{N}_i^{-T}$ and $\tilde{N}_i^{-1}$, respectively, we obtain

$$\left(M \hat{A}_i \tilde{N}_i^{-1} \right)^T \left\{ \sum_{j=1}^{\mathcal{N}} \pi_{ij} \left(M^{-T} \begin{bmatrix} \bar{P}_i & 0 \\ 0 & -\delta I \end{bmatrix} M^{-1} \right) \right\} \left(M \hat{A}_i \tilde{N}_i^{-1} \right)$$

$$- \left(M \hat{E} \tilde{N}_i^{-1} \right)^T \left(M^{-T} \begin{bmatrix} \bar{P}_i & 0 \\ 0 & -\delta I \end{bmatrix} M^{-1} \right) \left(M \hat{E} \tilde{N}_i^{-1} \right) < 0, \tag{10.45}$$

in which (10.29) is used. Now, let

$$P_i = M^{-T} \begin{bmatrix} \bar{P}_i & 0 \\ 0 & -\delta I \end{bmatrix} M^{-1}. \tag{10.46}$$

Then, by (10.41), (10.42) and (10.45), it can be verified that a set of matrices P_i, $i = 1, 2, \ldots, \mathcal{N}$, given in (10.46) satisfies (10.30) and (10.31). This completes the proof. $\qquad\square$

To show the applicability of Theorem 10.2, we provide the following example.

Example 10.1. Consider a discrete Markovian jump singular systems in (10.27) with

$$E = \begin{bmatrix} 1 & 0 & 0 \\ 0 & 1 & 0 \\ 0 & 0 & 0 \end{bmatrix}.$$

We suppose the system has two modes. For mode 1, the parameter of the system is given by

$$A_1 = \begin{bmatrix} 0.2 & -0.3 & 0 \\ 0.7 & -0.2 & -0.5 \\ 0.1 & 0 & 2 \end{bmatrix}.$$

For mode 2, the parameter of the system is given by

$$A_2 = \begin{bmatrix} -0.5 & 0.8 & 0.3 \\ -0.2 & 0.1 & -0.1 \\ 0 & -0.5 & 1 \end{bmatrix}.$$

The transition probability matrix is given by

$$\Pi = \begin{bmatrix} 0.2\ 0.8 \\ 0.4\ 0.6 \end{bmatrix}.$$

Then, it is found that the LMIs in (10.30) and (10.31) have a set of solutions as follows:

$$P_1 = \begin{bmatrix} 50.4543 & -5.0049 & -84.0057 \\ -5.0049 & 44.1601 & 145.9421 \\ -84.0057 & 145.9421 & -54.2049 \end{bmatrix},$$

$$P_2 = \begin{bmatrix} 39.0021 & -11.5701 & 16.6355 \\ -11.5701 & 56.5883 & -18.2548 \\ 16.6355 & -18.2548 & 4.9783 \end{bmatrix}.$$

Therefore, by Theorem 10.2, we have that the considered system is stochastically admissible.

10.3 Guaranteed Cost Control

In this section, we investigate the problem of guaranteed cost control for continuous uncertain Markovian jump singular systems. Attention will be focused on the design of state feedback controllers such that the closed-loop system is stochastically admissible and a quadratic cost function has an upper bound.

The class of continuous uncertain Markovian jump singular systems to be considered is described by

$$E\dot{x}(t) = [A(r_t) + \Delta A(r_t)]\,x(t) + [B(r_t) + \Delta B(r_t)]\,u(t), \qquad (10.47)$$

where $x(t) \in \mathbb{R}^n$ is the state of the system; $u(t) \in \mathbb{R}^m$ is the control input. The matrix $E \in \mathbb{R}^{n \times n}$ may be singular; we shall assume that $\text{rank}(E) = r \le n$; $\{r_t\}$ is a continuous-time Markovian process with right continuous trajectories taking values in a finite set given in (10.2) with transition probability matrix $\Pi = \{\pi_{ij}\}$ satisfying (10.3)–(10.5).

In (10.47), $A(r_t)$ and $B(r_t)$ are known real constant matrices representing the nominal system for each $r_t \in \mathcal{S}$, and $\Delta A(r_t)$ and $\Delta B(r_t)$ are unknown matrices representing time-invariant parameter uncertainties, and are assumed to be of the form

$$[\Delta A(r_t)\ \Delta B(r_t)] = M(r_t)\,F(r_t)\,[N_1(r_t)\ N_2(r_t)], \qquad (10.48)$$

where $M(r_t)$, $N_1(r_t)$ and $N_2(r_t)$ are known real constant matrices for all $r_t \in \mathcal{S}$, and $F(r_t)$, for all $r_t \in \mathcal{S}$, is the uncertain matrix satisfying

$$F(r_t)^T F(r_t) \le I. \tag{10.49}$$

The initial condition of the Markovian jump singular system in (10.47) is assumed to be

$$x(0) = \varphi(0).$$

Associated with system (10.47) is the following cost function:

$$J = \mathcal{E}\left\{\int_0^\infty \left[x(t)^T Q_1 x(t) + u(t)^T Q_2 u(t)\right] dt\right\}, \tag{10.50}$$

where $Q_1 > 0$ and $Q_2 > 0$ are given constant matrices.

Now, consider the following linear state feedback controller

$$u(t) = K(r_t) x(t), \quad K(r_t) \in \mathbb{R}^{m \times n}, \tag{10.51}$$

for $r_t \in \mathcal{S}$, where $K(r_t)$ is to be determined. Applying the controller in (6.5) to (6.1) results in the following closed-loop system:

$$E\dot{x}(t) = [A_c(r_t) + \Delta A_c(r_t)] x(t), \tag{10.52}$$

where

$$A_c(r_t) = A(r_t) + B(r_t) K(r_t), \tag{10.53}$$
$$\Delta A_c(r_t) = \Delta A(r_t) + \Delta B(r_t) K(r_t). \tag{10.54}$$

The guaranteed cost control problem to be addressed is formulated as follows: given two constant matrices $Q_1 > 0$ and $Q_2 > 0$, design a state feedback controller in (10.51) such that the closed-loop system (10.52) is stochastically admissible and the cost function in (10.50) has an upper bound for all uncertainties satisfying (10.48) and (10.49). In this case, (10.51) is said to be a guaranteed cost state feedback controller.

A sufficient condition for the solvability of the guaranteed cost control problem is given in the following theorem.

Theorem 10.3. *Consider the uncertain Markovian jump singular system in (10.47) and the cost function in (10.50). The guaranteed cost control problem is solvable if there exist matrices P_i, Y_i, scalars $\epsilon_i > 0$, $\delta_i > 0$, $i = 1, 2, \ldots, \mathcal{N}$, such that for $i = 1, 2, \ldots, \mathcal{N}$,*

$$P_i^T E^T = E P_i \ge 0, \tag{10.55}$$
$$P_i^T E^T \le \delta_i I, \tag{10.56}$$

$$\begin{bmatrix} \Psi_i + \epsilon_i M_i M_i^T & \Phi_i^T & P_i^T & Y_i^T & W_i \\ \Phi_i & -\epsilon_i I & 0 & 0 & 0 \\ P_i & 0 & -Q_1^{-1} & 0 & 0 \\ Y_i & 0 & 0 & -Q_2^{-1} & 0 \\ W_i^T & 0 & 0 & 0 & -J_i \end{bmatrix} < 0, \tag{10.57}$$

where

$$\Psi_i = \pi_{ii} P_i^T E^T + A_i P_i + B_i Y_i + (A_i P_i + B_i Y_i)^T, \qquad (10.58)$$

$$\Phi_i = N_{1i} P_i + N_{2i} Y_i, \qquad (10.59)$$

$$W_i = \left[\sqrt{\pi_{i1}} P_i^T \cdots \sqrt{\pi_{ii-1}} P_i^T \ \sqrt{\pi_{ii+1}} P_i^T \cdots \sqrt{\pi_{iN}} P_i^T \right], \qquad (10.60)$$

$$J_i = \mathrm{diag}\left(P_1^T + P_1 - \delta_1 I, \ldots, P_{i-1}^T + P_{i-1} - \delta_{i-1} I, \right.$$

$$\left. P_{i+1}^T + P_{i+1} - \delta_{i+1} I, \cdots, P_N^T + P_N - \delta_N I \right). \qquad (10.61)$$

In this case, we can assume that for $i = 1, 2, \ldots, N$, the matrix P_i is nonsingular (if this is not the case, then we can choose some $\theta_i \in (0, 1)$ such that $\hat{P}_i = P_i + \theta_i \tilde{P}_i$ is nonsingular and satisfies (10.55)–(10.57), in which $\tilde{P}_i$ is any nonsingular matrix satisfying $\tilde{P}_i^T E^T = E \tilde{P}_i \geq 0$ for $i = 1, 2, \ldots, N$), and a desired guaranteed cost state feedback controller can be chosen as

$$u(t) = Y_i P_i^{-1} x(t), \qquad (10.62)$$

for $i = 1, 2, \ldots, N$, and the corresponding cost function in (10.50) satisfies

$$J \leq \mathcal{E}\left\{ \varphi(0)^T E^T P_0^{-1} \varphi(0) \right\}. \qquad (10.63)$$

Proof. For $i = 1, 2, \ldots, N$, we denote

$$\tilde{P}_i = P_i^{-1}, \quad K_i = Y_i \tilde{P}.$$

Then, pre- and post-multiplying (10.57) by $\mathrm{diag}\left(\tilde{P}_i^T, I, I, I, I \right)$ and its transpose, respectively, we have

$$\begin{bmatrix} \tilde{\Psi}_i + \epsilon_i \tilde{P}_i^T M_i M_i^T \tilde{P}_i & N_{ci}^T & I & K_i^T & \tilde{W}_i \\ N_{ci} & -\epsilon_i I & 0 & 0 & 0 \\ I & 0 & -Q_1^{-1} & 0 & 0 \\ K_i & 0 & 0 & -Q_2^{-1} & 0 \\ \tilde{W}_i^T & 0 & 0 & 0 & -J_i \end{bmatrix} < 0, \qquad (10.64)$$

where

$$\tilde{\Psi}_i = \pi_{ii} E^T \tilde{P}_i + A_{ci}^T \tilde{P}_i + \tilde{P}_i^T A_{ci}, \qquad (10.65)$$

$$A_{ci} = A_i + B_i K_i, \qquad (10.66)$$

$$N_{ci} = N_{1i} + N_{2i} K_i, \qquad (10.67)$$

$$\tilde{W}_i = \left[\sqrt{\pi_{i1}} I, \cdots, \sqrt{\pi_{ii-1}} I, \sqrt{\pi_{ii+1}} I, \cdots, \sqrt{\pi_{iN}} I \right]. \qquad (10.68)$$

Note that for $j = 1, 2, \ldots, i-1, i+1, \ldots, N$,

$$\delta_j^{-1} P_j^T P_j \geq P_j^T + P_j - \delta_j I. \qquad (10.69)$$

Then, it follows from (10.64) and (10.69) that

$$\begin{bmatrix} \tilde{\Psi}_i + \epsilon_i \tilde{P}_i^T M_i M_i^T \tilde{P}_i & N_{ci}^T & I & K_i^T & \tilde{W}_i \\ N_{ci} & -\epsilon_i I & 0 & 0 & 0 \\ I & 0 & -Q_1^{-1} & 0 & 0 \\ K_i & 0 & 0 & -Q_2^{-1} & 0 \\ \tilde{W}_i^T & 0 & 0 & 0 & -\tilde{J}_i \end{bmatrix} < 0, \qquad (10.70)$$

where

$$\tilde{J}_i = \left[\delta_1^{-1} P_1^T P_1, \cdots, \delta_{i-1}^{-1} P_{i-1}^T P_{i-1}, \delta_{i+1}^{-1} P_{i+1}^T P_{i+1}, \cdots, \delta_{\mathcal{N}}^{-1} P_{\mathcal{N}}^T P_{\mathcal{N}} \right].$$

Applying Schur complement to (10.70) gives

$$\sum_{j \neq i}^{\mathcal{N}} \delta_j \pi_{ij} \tilde{P}_j^T \tilde{P}_j + \pi_{ii} E^T \tilde{P}_i + A_{ci}^T \tilde{P}_i + \tilde{P}_i^T A_{ci}$$
$$+ \epsilon_i \tilde{P}_i^T M_i M_i^T \tilde{P}_i + \epsilon_i^{-1} N_{ci}^T N_{ci} + Q_1 + K_i^T Q_2 K_i < 0. \qquad (10.71)$$

For $j = 1, 2, \ldots, i-1, i+1 \ldots, \mathcal{N}$, pre- and post-multiplying (10.56) by $\tilde{P}_i^T$ and $\tilde{P}_i$, respectively, we have that for $i = 1, 2, \ldots, \mathcal{N}$,

$$\delta_j \tilde{P}_i^T \tilde{P}_i \geq E^T \tilde{P}_i.$$

This together with (10.71) implies

$$\sum_{j=1}^{\mathcal{N}} \pi_{ij} E^T \tilde{P}_j + A_{ci}^T \tilde{P}_i + \tilde{P}_i^T A_{ci}$$
$$+ \epsilon_i \tilde{P}_i^T M_i M_i^T \tilde{P}_i + \epsilon_i^{-1} N_{ci}^T N_{ci} + Q_1 + K_i^T Q_2 K_i < 0. \qquad (10.72)$$

Observe that

$$\Delta A_{ci} = \Delta A_i + \Delta B_i K_i = M_i F_i N_{ci}.$$

Then, we have that for any $\epsilon_i > 0$,

$$\Delta A_{ci}^T \tilde{P}_i + \tilde{P}_i^T \Delta A_{ci} \leq \epsilon_i \tilde{P}_i^T M_i M_i^T \tilde{P}_i + \epsilon_i^{-1} N_{ci}^T N_{ci}.$$

With this and (10.72), it is easy to obtain

$$\sum_{j=1}^{\mathcal{N}} \pi_{ij} E^T \tilde{P}_j + (A_{ci} + \Delta A_{ci})^T \tilde{P}_i + \tilde{P}_i^T (A_{ci} + \Delta A_{ci}) + Q_1 + K_i^T Q_2 K_i < 0.$$
$$(10.73)$$

Therefore,

$$\sum_{j=1}^{\mathcal{N}} \pi_{ij} E^T \tilde{P}_j + (A_{ci} + \Delta A_{ci})^T \tilde{P}_i + \tilde{P}_i^T (A_{ci} + \Delta A_{ci}) < 0. \qquad (10.74)$$

For $i = 1, 2, \ldots, \mathcal{N}$, pre- and post-multiplying (10.55) by $\tilde{P}_i^T$ and $\tilde{P}_i$, respectively, provide

$$E^T \tilde{P}_i = \tilde{P}_i^T E \geq 0.$$

Noting this inequality together with (10.74) and then using Theorem 10.1, we have that the closed-loop system (10.52) with the controller in (10.62) is stochastically admissible. To show the performance in (10.63), we define

$$V\left(x\left(t\right),i\right) = x\left(t\right)^T E^T \tilde{P}_i x\left(t\right),$$

and let $\mathcal{A}$ be the weak infinitesimal generator of the random process $\{x\left(t\right), r_t\}$. Then, for each $r_t = i$, $i \in \mathcal{S}$, we have

$$\mathcal{A}V(x\left(t\right),i) = x\left(t\right)^T \left[\sum_{j=1}^N \pi_{ij} E^T \tilde{P}_j + \left(A_{ci} + \Delta A_{ci}\right)^T \tilde{P}_i + \tilde{P}_i^T \left(A_{ci} + \Delta A_{ci}\right) \right] x\left(t\right).$$

Considering this and (10.74), we obtain that for each $i \in \mathcal{S}$,

$$\mathcal{A}V(x\left(t\right),i) \le -x\left(t\right)^T \left(Q_1 + K_i^T Q_2 K_i\right) x\left(t\right). \tag{10.75}$$

Now applying Dynkin's formula to (10.75), we have that for each $r_t = i$, $i \in \mathcal{S}$, and $T > 0$,

$$\mathcal{E}\left\{V(x\left(T\right),r_T)\right\} - \mathcal{E}\left\{V(x\left(0\right),r_0)\right\}$$
$$= \mathcal{E}\left\{\int_0^T \mathcal{A}V(x\left(t\right),r_t)dt\right\}$$
$$\le -\mathcal{E}\left\{\int_0^T x\left(t\right)^T \left(Q_1 + K_{r_t}^T Q_2 K_{r_t}\right) x\left(t\right)dt\right\}.$$

That is, for any $T > 0$,

$$\mathcal{E}\left\{\int_0^T x\left(t\right)^T \left(Q_1 + K_{r_t}^T Q_2 K_{r_t}\right) x\left(t\right)dt\right\} \le \mathcal{E}\left\{\varphi\left(0\right)^T E^T \tilde{P}_0 \varphi\left(0\right)\right\}.$$

Therefore, the performance in (10.63) is satisfied. This completes the proof.□

10.4 H_∞ Control

In this section, we study the H_∞ control problem for continuous Markovian jump singular systems via state feedback controllers. The class of linear continuous Markovian jump singular systems to be considered is described by

$$E\dot{x}(t) = A\left(r_t\right)x(t) + B_1\left(r_t\right)u(t) + D_1\left(r_t\right)\omega(t), \tag{10.76}$$
$$z(t) = C\left(r_t\right)x(t) + B_2\left(r_t\right)u(t) + D_2\left(r_t\right)\omega(t), \tag{10.77}$$

where $x(t) \in \mathbb{R}^n$ is the state; $u(t) \in \mathbb{R}^m$ is the control input; $w(t) \in \mathbb{R}^p$ is the disturbance input which belongs to $\mathcal{L}_2[0, \infty)$, and $z(t) \in \mathbb{R}^s$ is the controlled output. The matrix $E \in \mathbb{R}^{n \times n}$ may be singular; we shall assume that $\text{rank}(E) = r \leq n$; $\{r_t\}$ is a continuous-time Markovian process with right continuous trajectories taking values in a finite set in (10.2) with transition probability matrix $\Pi = \{\pi_{ij}\}$ satisfies (10.3)–(10.5). For $r_t \in \mathcal{S}$, $A(r_t)$, $B_1(r_t)$, $B_2(r_t)$, $C(r_t)$, $D_1(r_t)$ and $D_2(r_t)$ are known real constant matrices with appropriate dimensions.

Now, we consider the following linear state feedback controller

$$u(t) = K(r_t) x(t), \quad K(r_t) \in \mathbb{R}^{m \times n}, \tag{10.78}$$

for $r_t \in \mathcal{S}$, where $K(r_t)$ is to be determined. Applying the controller in (10.78) to (10.76) and (10.77), we have the following closed-loop system:

$$E\dot{x}(t) = A_c(r_t) x(t) + D_1(r_t) w(t), \tag{10.79}$$
$$z(t) = C_c(r_t) x(t) + D_2(r_t) w(t), \tag{10.80}$$

where

$$A_c(r_t) = A(r_t) + B_1(r_t) K(r_t), \tag{10.81}$$
$$C_c(r_t) = C(r_t) + B_2(r_t) K(r_t). \tag{10.82}$$

The H_∞ control problem to be addressed is formulated as follows: given a scalar $\gamma > 0$, design a state feedback controller in (10.78) such that the closed-loop system in (10.79) and (10.80) is stochastically admissible and the following H_∞ performance

$$|z|_{E_2} < \gamma |w|_2, \tag{10.83}$$

holds under zero-initial conditions for any nonzero $w \in \mathcal{L}_2[0, \infty)$, where $|\cdot|_2$ denotes the standard norm in $\mathcal{L}_2[0, \infty)$ and

$$|\cdot|_{E_2} \triangleq \mathcal{E}\{|\cdot|_2\}.$$

Now, we present a sufficient condition for the solvability of the H_∞ control problem in the following theorem.

Theorem 10.4. *Given a scalar $\gamma > 0$ and the continuous Markovian jump singular system in (10.76) and (10.77). The H_∞ control problem is solvable if there exist matrices P_i, Y_i, scalars $\delta_i > 0$, $i = 1, 2, \ldots, \mathcal{N}$, such that for $i = 1, 2, \ldots, \mathcal{N}$,*

$$P_i^T E^T = E P_i \geq 0, \tag{10.84}$$
$$P_i^T E^T \leq \delta_i I, \tag{10.85}$$

$$\begin{bmatrix} \Lambda_i & D_{1i} & \Xi_i^T & W_i \\ D_{1i}^T & -\gamma^2 I & D_{2i}^T & 0 \\ \Xi_i & D_{2i} & -I & 0 \\ W_i^T & 0 & 0 & -J_i \end{bmatrix} < 0, \tag{10.86}$$

where W_i and J_i are given in (10.60) and (10.61), respectively, and

$$\Lambda_i = \pi_{ii} P_i^T E^T + A_i P_i + B_{1i} Y_i + (A_i P_i + B_{1i} Y_i)^T,$$
$$\Xi_i = C_i P_i + B_{2i} Y_i.$$

In this case, we can assume that for $i = 1, 2, \ldots, \mathcal{N}$, the matrix P_i is nonsingular (if this is not the case, then we can choose some $\theta_i \in (0, 1)$ such that $\hat{P}_i = P_i + \theta_i \tilde{P}_i$ is nonsingular and satisfies (10.84)–(10.86), in which $\tilde{P}_i$ is any nonsingular matrix satisfying $\tilde{P}_i^T E^T = E \tilde{P}_i \geq 0$ for $i = 1, 2, \ldots, \mathcal{N}$), and a desired state feedback controller can be chosen as

$$u(t) = Y_i P_i^{-1} x(t). \tag{10.87}$$

Proof. For $i = 1, 2, \ldots, \mathcal{N}$, we denote

$$\tilde{P}_i = P_i^{-1}, \quad K_i = Y_i \tilde{P}. \tag{10.88}$$

Then, pre- and post-multiplying (10.86) by $\text{diag}(\tilde{P}_i^T, I, I, I)$ and its transpose, respectively, we have

$$\begin{bmatrix} \pi_{ii} E^T \tilde{P}_i + A_{ci}^T \tilde{P}_i + \tilde{P}_i^T A_{ci} & \tilde{P}_i^T D_{1i} & C_{ci}^T & \tilde{W}_i \\ D_{1i}^T \tilde{P}_i & -\gamma^2 I & D_{2i}^T & 0 \\ C_{ci} & D_{2i} & -I & 0 \\ \tilde{W}_i^T & 0 & 0 & -J_i \end{bmatrix} < 0, \tag{10.89}$$

where $\tilde{W}$ is given in (10.68), and A_{ci} and C_{ci} are given in (10.81) and (10.82), respectively, with K_i in (10.88). Applying Schur complement to (10.89) gives

$$\begin{bmatrix} \sum_{j \neq i}^{\mathcal{N}} \delta_j \pi_{ij} \tilde{P}_j^T \tilde{P}_j + \pi_{ii} E^T \tilde{P}_i + A_{ci}^T \tilde{P}_i + \tilde{P}_i^T A_{ci} & \tilde{P}_i^T D_{1i} \\ D_{1i}^T \tilde{P}_i & -\gamma^2 I \end{bmatrix}$$
$$+ \begin{bmatrix} C_{ci}^T \\ D_{2i}^T \end{bmatrix} \begin{bmatrix} C_{ci}^T \\ D_{2i}^T \end{bmatrix}^T < 0.$$

Now, using this inequality together with (10.84) and (10.85), and then following a similar line as in the proof of Theorem 10.3, we can show that

$$\Upsilon_i = \begin{bmatrix} \sum_{j=1}^{\mathcal{N}} \pi_{ij} E^T \tilde{P}_j + A_{ci}^T \tilde{P}_i + \tilde{P}_i^T A_{ci} & \tilde{P}_i^T D_{1i} \\ D_{1i}^T \tilde{P}_i & -\gamma^2 I \end{bmatrix}$$
$$+ \begin{bmatrix} C_{ci}^T \\ D_{2i}^T \end{bmatrix} \begin{bmatrix} C_{ci}^T \\ D_{2i}^T \end{bmatrix}^T < 0, \tag{10.90}$$

and

$$E^T \tilde{P}_i = \tilde{P}_i^T E \geq 0. \tag{10.91}$$

By (10.90), it is easy to see that

$$\sum_{j=1}^{\mathcal{N}} \pi_{ij} E^T \tilde{P}_j + A_{ci}^T \tilde{P}_i + \tilde{P}_i^T A_{ci} < 0.$$

Noting this and (10.91), and then using Theorem 10.1, we have that the closed-loop system (10.79) with the controller in (10.87) is stochastically admissible. To show the H_∞ performance, we set

$$V(x(t)) = x(t)^T E^T \tilde{P}_i x(t),$$

and define

$$J(\mathcal{T}) = \mathcal{E} \left\{ \int_0^{\mathcal{T}} \left[z(t)^T z(t) - \gamma^2 \omega(t)^T \omega(t) \right] dt \right\},$$

where $\mathcal{T} > 0$. Also, let $\mathcal{A}$ be the weak infinitesimal generator of the random process $\{x(t), r_t\}$. Then, under the initial condition $x(0) = 0$, it can be shown that

$$J(\mathcal{T}) = \mathcal{E} \left\{ \int_0^{\mathcal{T}} \left[z(t)^T z(t) - \gamma^2 \omega(t)^T \omega(t) + \mathcal{A}V(x_t, r_t, t) \right] dt \right\} - \mathcal{E} \left\{ V(x_{\mathcal{T}}, r_{\mathcal{T}}) \right\}$$

$$\leq \mathcal{E} \left\{ \int_0^{\mathcal{T}} \left[z(t)^T z(t) - \gamma^2 \omega(t)^T \omega(t) + \mathcal{A}V(x_t, r_t, t) \right] dt \right\}$$

$$= \mathcal{E} \left\{ \int_0^{\mathcal{T}} \eta(t)^T \Upsilon(r_t) \eta(t) \, dt \right\}, \tag{10.92}$$

where

$$\eta(t) = \left[x(t)^T \quad \omega(t)^T \right]^T,$$

and $\Upsilon(r_t)$ is denoted as Υ_i when $r_t = i$, $i \in \mathcal{S}$ which is given in (10.90). Then, it follows from (10.90) and (10.92) that for any $\mathcal{T} > 0$,

$$J(\mathcal{T}) < 0.$$

Therefore, the H_∞ performance in (10.83) is satisfied. This completes the proof.   □

10.5 Conclusion

This chapter has studied the problems of stability analysis and control of Markovian jump singular systems. In the stability analysis, we have provided necessary and sufficient conditions for stochastic admissibility of a Markovian jump singular system in the continuous and discrete cases, respectively. Then, the problem of guaranteed cost control for continuous uncertain Markovian jump singular systems has been considered and a sufficient condition for the solvability of this problem has been given. Finally, we have addressed the H_∞ control problem for continuous Markovian jump singular systems. State feedback controllers have been designed such that the closed-loop system is stochastically admissible and a prescribed H_∞ performance level is satisfied.

A

List of Symbols and Acronyms

Symbols

$\square$	end of proof
$\triangleleft$	end of remark
$\diamond$	end of example
$\triangleq$	' is defined as '
$\in$	' belongs to '
$\mathbb{C}$	field of complex numbers
$\mathbb{R}$	field of real numbers
$\mathbb{R}^n$	space of n-dimensional real vectors
$\mathbb{R}^{n \times m}$	space of $n \times m$ real matrices
$\mathrm{rank}(\cdot)$	rank of a matrix
$\det(\cdot)$	determinant of a matrix
$\deg(\cdot)$	degree of a polynomial
$\lambda_{\min}(\cdot)$	minimum eigenvalue of a real symmetric matrix
$\lambda_{\max}(\cdot)$	maximum eigenvalue of a real symmetric matrix
$\mathrm{Re}(\cdot)$	real part of a complex number
$|\cdot|$	Euclidean vector norm
$\|\cdot\|$	Euclidean matrix norm (spectral norm)
$\alpha(\cdot,\cdot)$	generalized spectral abscissa of a matrix
$\alpha(\cdot)$	spectral abscissa of a matrix
$\rho(\cdot,\cdot)$	generalized spectral radius of a matrix

$\rho(\cdot)$	spectral radius of a matrix
I	identity matrix
I_n	$n \times n$ identity matrix
X^T	transpose of matrix X
X^*	conjugate transpose of matrix X
X^{-1}	inverse of matrix X
X^+	Moore-Penrose inverse of matrix X
$X^\perp$	full row rank matrix satisfying $X^\perp X = 0$ and $X^\perp X^{\perp T} > 0$
$\mathrm{diag}(X_1, \ldots, X_m)$	block diagonal matrix with blocks $X_1, \ldots, X_m$
$X > (<)Y$	$X - Y$ is real symmetric and positive (negative) definite
$X \geq (\leq)Y$	$X - Y$ is real symmetric and positive (negative) semi-definite
$\mathcal{L}_2[0, \infty)$	space of square integrable functions on $[0, \infty)$
$l_2[0, \infty)$	space of square summable infinite vector sequences over $[0, \infty)$
$\|\cdot\|_2$	$\mathcal{L}_2$-norm: $\left(\int_0^\infty |\cdot|^2 dt\right)^{1/2}$ (continuous case)
	l_2-norm: $\left(\sum_0^\infty |\cdot|^2\right)^{1/2}$ (discrete case)
$\|\cdot\|_{E_2}$	$\mathcal{E}\{|\cdot|_2\}$
$\|G\|_\infty$	H_∞ norm of transfer function G:
	$\sup_{\omega \in [0,\infty)} \|G(j\omega)\|$ (continuous case)
	$\sup_{\omega \in [0,2\pi)} \|G(e^{j\omega})\|$ (discrete case)
$\mathcal{E}\{\cdot\}$	mathematical expectation operator

Acronyms

BMI	bilinear matrix inequality
ESPR	extended strictly positive real
LMI	linear matrix inequality
PR	positive real
SPR	strictly positive real

References

1. A. Ailon. Controllability of generalized linear time-invariant systems. *IEEE Trans. Automat. Control*, 32:429–432, 1987.
2. A. Ailon. On the design of output-feedback for finite and infinite pole assignment in singular systems with application to the control problem of constrained robots. *Circuit, Syst, Sig. Process.*, 13:525–544, 1994.
3. B. D. O. Anderson and J. B. Moore. *Optimal Filtering.* Prentice-Hall, Englewood Cliffs, NJ, 1979.
4. B. D. O. Anderson and J. B. Moore. *Linear Optimal Control.* Prentice-Hall, NJ: Englewood Cliffs, 1990.
5. B. D. O. Anderson and S. Vongpanitlerd. *Network Analysis and Synthesis: A Modern Systems Theory Approach.* Upper Saddle River, NJ: Prentice-Hall, 1973.
6. J. D. Aplevich. *Implicit Linear Systems.* Berlin: Springer-Verlag, 1991.
7. A. Banaszuk, M. Kociecki, and F. L. Lewis. Kalman decomposition for implicit linear systems. *IEEE Trans. Automat. Control*, 37:1509–1514, 1992.
8. B. R. Barmish. Necessary and sufficient conditions for quadratic stabilizability of an uncertain systems. *J. Optim. Theory Appl.*, 46:399–408, 1985.
9. U. Başer and J. M. Schumacher. The equivalence structure of descriptor representations of systems with possibly inconsistent initial conditions. *Linear Algebra Appl.*, 318:53–77, 2000.
10. T. Beellen and P. Vandooren. A numerical-method for deadbeat control of generalized state-space systems. *Systems & Control Lett.*, 10:225–233, 1988.
11. D. J. Bender. Lyapunov-like equations and reachability observability gramians for descriptor systems. *IEEE Trans. Automat. Control*, 32:343–348, 1987.
12. D. J. Bender and A. J. Laub. The linear-quadratic optimal regulator for descriptor systems. *IEEE Trans. Automat. Control*, 32:672–687, 1987.
13. D. S. Bernstein and W. M. Haddad. Steady-state Kalman filtering with an H_∞ error bound. *Systems & Control Lett.*, 12:9–16, 1989.
14. S. Bittanti and F. A. Cuzzola. Continuous-time periodic H_∞ filtering via LMI. *European Journal of Control*, 7:2–16, 2001.
15. S. Boyd, L. El Ghaoui, E. Feron, and V. Balakrishnan. *Linear Matrix Inequalities in System and Control Theory.* SIAM Studies in Applied Mathematics. SIAM, Philadelphia, Pennsylvania, 1994.

16. A. Bunse-Gerstner, V. Mehrmann, and N. K. Nichols. Regularization of descriptor systems by derivative and proportional state feedback. *SIAM J. Matrix Anal. Appl.*, 13:46–67, 1992.

17. A. Bunse-Gerstner, V. Mehrmann, and N. K. Nichols. Regularization of descriptor systems by output feedback. *IEEE Trans. Automat. Control*, 39:1742–1748, 1994.

18. R. Byers, P. Kunkel, and V. Mehrmann. Regularization of linear descriptor systems with variable coefficients. *SIAM J. Control Optim.*, 35:117–133, 1997.

19. G. D. Byrne and P. R. Ponzi. Differential-algebraic systems, their applications and solutions. *Comput. Chem. Engng.*, 12:377–382, 1988.

20. S. L. Campbell. *Singular Systems of Differential Equations*. San Francisco: Pitman, 1980.

21. S. L. Campbell. *Singular Systems of Differential Equations II*. San Francisco: Pitman, 1982.

22. S. L. Campbell. One canonical form for higher-index linear time-varying singular systems. *Circuit, Syst, Sig. Process.*, 2:311–326, 1983.

23. S. L. Campbell and E. Griepentrog. Solvability of general differential algebraic equations. *SIAM J. Sci. Comput.*, 16:257–270, 1995.

24. S.-H. Chen. Robust D-stability analysis for linear discrete-time singular systems with structured parameter uncertainties and delayed perturbations. *Proc. Inst. Mech. Eng. Part I-J Syst Control Eng.*, 217:1–5, 2003.

25. J. H. Chou. Stability robustness of linear state space models with structured perturbations. *Syst. Contr. Lett.*, 15:207–210, 1990.

26. M. A. Christodoulou and C. Isik. Feedback-control for nonlinear singular systems. *Int. J. Control*, 51:487–494, 1990.

27. M. A. Christodoulou and P. N. Paraskevopoulos. Solvability, controllability, and observability of singular systems. *J. Optim. Theory Appl.*, 45:53–72, 1985.

28. D. L. Chu, H. C. Chan, and D. W. C. Ho. Regularization of singular systems by derivative and proportional output feedback. *SIAM J. Matrix Anal. Appl.*, 19:21–38, 1998.

29. D. L. Chu and D. W. C. Ho. Necessary and sufficient conditions for the output feedback regularization of descriptor systems. *IEEE Trans. Automat. Control*, 44:405–412, 1999.

30. D. Cobb. Feedback and pole placement in descriptor variable systems. *Int. J. Control*, 33:1135–1146, 1981.

31. D. Cobb. Descriptor variable systems and optimal state regulation. *IEEE Trans. Automat. Control*, 28:601–611, 1983.

32. D. Cobb. Controllability, observability, and duality in singular systems. *IEEE Trans. Automat. Control*, 29:1076–1082, 1984.

33. G. Conte and A. Perdon. Generalized state-space realizations of non-proper rational transfer-functions. *Systems & Control Lett.*, 1:270–276, 1982.

34. L. Dai. Observers for discrete singular systems. *IEEE Trans. Automat. Control*, 33:187–191, 1988.

35. L. Dai. Filtering and LQG problems for discrete-time stochastic singular systems. *IEEE Trans. Automat. Control*, 34:1105C1108, 1989.

36. L. Dai. *Singular Control Systems*. Berlin: Springer-Verlag, 1989.

37. M. Darouach, M. Zasadzinski, and M. Hayar. Reduced-order observer design for descriptor systems with unknown inputs. *IEEE Trans. Automat. Control*, 41:1068–1072, 1996.

38. M. Darouach, M. Zasadzinski, and D. Mehdi. State estimation of stochastic singular linear-systems. *Int. J. Systems Sci.*, 24:345–354, 1993.

39. C. E. de Souza and M. D. Fragoso. H_∞ control for linear systems with Markovian jumping parameters. *Control Theory and Advance Technology*, 9:457–466, 1993.

40. C. A. Desoer and M. Vidyasagar. *Feedback Systems: Input-Output Properties.* New York-London: Academic Press, 1975.

41. J. C. Doyle, K. Glover, P. P. Khargonekar, and B. A. Francis. State-space solutions to standard H_2 and H_∞ control problems. *IEEE Trans. Automat. Control*, 34:831–847, 1989.

42. V. Dragan and T. Morozan. Stability and robust stabilization to linear stochastic systems described by differential equations with Markov jumping and multiplicative white noise. *Stochastic Anal. Appl.*, 20:33–92, 2002.

43. G. R. Duan and R. J. Patton. Eigenstructure assignment in descriptor systems via state feedback–a new complete parametric approach. *Int. J. Systems Sci.*, 29:167–178, 1998.

44. L. Dugard and E. I. Verriest. *Stability and Control of Time-delay Systems.* London: Springer-Verlag, 1998.

45. S. H. Esfahani, S. O. R. Moheimani, and I. R. Petersen. LMI approach suboptimal quadratic guaranteed cost control for uncertain time-delay systems. *IEE Proc. Control Theory Appl.*, 145:491–498, 1998.

46. S. H. Esfahani and I. R. Petersen. An LMI approach to the output-feedback guaranteed cost control for uncertain time-delay systems. In *Proc. 37th Conf. Decision Control*, pages 1358–1363, Tampa, Florida, USA, December, 1998.

47. C.-H. Fang and F.-R. Chang. A strongly observable and controllable realization of descriptor systems. *Control Theory Adv. Tech.*, 6:133–141, 1990.

48. C.-H. Fang and F.-R. Chang. Analysis of stability robustness for generalized state-space systems with structured perturbations. *Systems & Control Lett.*, 21:109–114, 1993.

49. C.-H. Fang, L. Lee, and F.-R. Chang. Robust control analysis and design for discrete-time singular systems. *Automatica*, 30:1741–1750, 1994.

50. X. Feng, K. A. Loparo, Y. Ji, and H. J. Chizeck. Stochastic stability properties of jump linear systems. *IEEE Trans. Automat. Control*, 37:38–53, 1992.

51. L. R. Fletcher. Regularizability of descriptor systems. *Int. J. Systems Sci.*, 17:843–847, 1986.

52. E. Fridman and U. Shaked. H_∞-control of linear state-delay descriptor systems: an LMI approach. *Linear Algebra Appl.*, 351/352:271–302, 2002.

53. M. Fu. Interpolation approach to H_∞ optimal estimation and its interconnection to loop transfer recovery. *Systems & Control Lett.*, 17:29–36, 1991.

54. M. Fukuda and M. Kojima. Branch-and-cut algorithms for the bilinear matrix inequality eigenvalue problem. *Comput. Optim. Appl.*, 19:79–105, 2001.

55. P. Gahinet and P. Apkarian. A linear matrix inequality approach to H_∞ control. *Int. J. Robust & Nonlinear Control*, 4:421–448, 1994.

56. Z. Gao. PD observer parametrization design for descriptor systems. *J. Franklin Inst.*, 342:551–564, 2005.

57. G. Garcia, J. Bernussou, and D. Arzelier. Robust stabilization of discrete-time linear systems with norm-bounded time-varying uncertainty. *Systems & Control Lett.*, 22:327–339, 1994.

58. T. Geerts. Solvability conditions, consistency, and weak consistency for linear differential-algebraic equations and time-invariant singular systems: The general case. *Linear Algebra Appl.*, 181:111–130, 1993.

59. T. Geerts. Stability concepts for general continuous-time implicit systems-definitions, hautus tests and lyapunov criteria. *Int. J. Systems Sci.*, 26:481–498, 1995.

60. E. Gershon, D. J. N. Limebeer, U. Shaked, and I. Yaesh. Robust H_∞ filtering of stationary continuous-time linear systems with stochastic uncertainties. *IEEE Trans. Automat. Control*, 46:1788–1793, 2001.

61. H. Glusingluerssen. Feedback canonical form for singular systems. *Int. J. Control*, 52:347–376, 1990.

62. H. Glusingluerssen and D. Hinrichsen. A Jordan control canonical form for singular systems. *Int. J. Control*, 48:1769–1785, 1988.

63. M. S. Goodwin. Exact pole assignment with regularity by output feedback in descriptor systems. II. *Int. J. Control*, 62:413–441, 1995.

64. M. S. Goodwin and L. R. Fletcher. Exact pole assignment with regularity by output feedback in descriptor systems. I. *Int. J. Control*, 62:379–411, 1995.

65. K. M. Grigoriadis. Optimal H_∞ model reduction via linear matrix inequalities: continuous- and discrete-time cases. *Systems & Control Lett.*, 26:321–333, 1995.

66. K. M. Grigoriadis and J. T. Watson. Reduced-order H_∞ and L_2-L_∞ filtering via linear matrix inequalities. *IEEE Trans. Aerospace and Electronic Systems*, 33:1326–1338, 1997.

67. J. Grimm. Realization and canonicity for implicit systems. *SIAM J. Control Optim.*, 26:1331–1347, 1988.

68. Z.-H. Guan and J. Lam. Robust stability of composite singular and impulsive interval uncertain dynamic systems with time delay. 32:1297–1307, 2001.

69. L. Guo and M. Malabre. Robust H_∞ control for descriptor systems with nonlinear uncertainties. *Int. J. Control*, 76:1254–1262, 2003.

70. W. M. Haddad and D. S. Bernstein. Robust stabilization with positive real uncertainty: beyond the small gain theorem. *Systems & Control Lett.*, 17:191–208, 1991.

71. W. M. Haddad and D. S. Bernstein. Explicit construction of quadratic Lyapunov functions for the small gain, positivity, circle, and Popov theorems and their application to robust stability. part II: discrete-time theory. *Int. J. Robust & Nonlinear Control*, 4:249–265, 1994.

72. J. K. Hale. *Theory of Functional Differential Equations*. New York: Springer-Verlag, 1977.

73. G. Hayton, P. Fretwell, and A. Pugh. Fundamental equivalence of generalized state-space systems. *IEEE Trans. Automat. Control*, 31:431–439, 1986.

74. U. Helmke and M. A. Shayman. A canonical form for controllable singular systems. *Systems & Control Lett.*, 12:111–122, 1989.

75. H. Hemami and B. F. Wyman. Modeling and control of constrained dynamic systems with application to biped locomotion in the frontal plane. *IEEE Trans. Automat. Control*, 24:526–535, 1979.

76. D. Hinrichsen, W. Manthey, and U. Helmke. Minimal partial realization by descriptor systems. *Linear Algebra Appl.*, 326:45–84, 2001.

77. D. Hinrichsen and J. Ohalloran. A complete characterization of orbit closures of controllable singular systems under restricted system equivalence. *SIAM J. Control Optim.*, 28:602–623, 1990.

78. M. Hou. Controllability and elimination of impulsive modes in descriptor systems. *IEEE Trans. Automat. Control*, 49:1723–1727, 2004.

79. M. Hou and P. C. Muller. Observer design for descriptor systems. *IEEE Trans. Automat. Control*, 44:164–169, 1999.

80. K.-L. Hsiung and L. Lee. Lyapunov inequality and bounded real lemma for discrete-time descriptor systems. *IEE Proc.-Control Theory Appl.*, 146:327–331, 1999.

81. J.-C. Huang, H.-S. Wang, and F.-R. Chang. Robust H_∞ control for uncertain linear time-invariant descriptor systems. *IEE Proc.-Control Theory Appl.*, 147:648–654, 2000.

82. J. Y. Ishihara and M. H. Terra. On the Lyapunov theorem for singular systems. *IEEE Trans. Automat. Control*, 47:1926–1930, 2002.

83. J. Y. Ishihara, M. H. Terra, and R. M. Sales. The full information and state feedback H_2 optimal controllers for descriptor systems. *Automatica*, 39:391–402, 2003.

84. T. Iwasaki and R. E. Skelton. All controllers for the general H_∞ control problems: LMI existence conditions and state space formulas. *Automatica*, 30:1307–1317, 1994.

85. Y. Ji and H. J. Chizeck. Jump linear quadratic Gaussian control: steady-state solution and testable conditions. *Control Theory Adv. Tech.*, 6:289–319, 1990.

86. S. H. Jin and J. B. Park. Robust H_∞ filtering for polytopic uncertain systems via convex optimisation. *IEE Proc.-Control Theory Appl.*, 148:55–59, 2001.

87. T. Kaczorek. Stabilization of linear descriptor systems by state-feedback controllers. *Appl. Math. Comput. Sci.*, 6:27–32, 1996.

88. P. P. Khargonekar, I. R. Petersen, and K. Zhou. Robust stabilization of uncertain linear systems: quadratic stabilizability and H_∞ control theory. *IEEE Trans. Automat. Control*, 35:356–361, 1990.

89. D. Koenig and S. Mammar. Design of proportional-integral observer for unknown input descriptor systems. *IEEE Trans. Automat. Control*, 47:2057–2062, 2002.

90. V. B. Kolmanovskii and A. D. Myshkis. *Applied Theory of Functional Differential Equations*. Dordrecht: Kluwer Academic Publishers, 1992.

91. F. N. Koumboulis and P. N. Paraskevopoulos. On the pole assignment of generalized state-space systems via state feedback. *IEE Proc.-Control Theory Appl.*, 139:106–108, 1992.

92. F. N. Koumboulisa and P. N. Paraskevopoulos. On the generic controllability of continuous generalized state-space systems. *Automatica*, 29:527–530, 1993.

93. M. Krstić, I. Kanellakopoulos, and P. V. Kokotovic. *Nonlinear and Adaptive Control Design*. New York: John Wiley Sons Inc., 1995.

94. M. Kuijper and J. M. Schumacher. Realization of autoregressive equations in pencil and descriptor form. *SIAM J. Control Optim.*, 28:1162–1189, 1990.

95. A. Kumar and P. Daoutidis. Feedback control of nonlinear differential-algebraic equation systems. *AIChE Journal*, 41:619–636, 1995.

96. A. Kumar and P. Daoutidis. State-space realizations of linear differential-algebraic-equation systems with control-dependent state space. *IEEE Trans. Automat. Control*, 41:269–274, 1996.

97. A. Kumar and P. Daoutidis. *Control of Nonlinear Differential Algebraic Equation Systems*. Boca Raton : Chapman & Hall/CRC, 1999.

98. P. Kunkel and V. Mehrmann. The linear quadratic optimal control problem for linear descriptor systems with variable coefficients. *Maths. Contr. Sig. Sys.*, 10:247–264, 1997.

99. Y.-C. Kuo, W.-W. Lin, and S.-F. Xu. Regularization of linear discrete-time periodic descriptor systems by derivative and proportional state feedback. *SIAM J. Matrix Anal. Appl.*, 25:1046–1073, 2003.

100. G. A. Kurina and R. März. On linear-quadratic optimal control problems for time-varying descriptor systems. *SIAM J. Control Optim.*, 42:2062–2077, 2004.

101. P. Lancaster and M. Tismenetsky. *The Theory of Matrices.* 2nd edition. New York: Academic Press, 1985.

102. G. Lebret and J. J. Loiseau. Proportional and proportional-derivative canonical-forms for descriptor systems with outputs. *Automatica*, 30:847–864, 1994.

103. J. H. Lee, S. W. Kim, and W. H. Kwon. Memoryless H_∞ controllers for state delayed systems. *IEEE Trans. Automat. Control*, 39:159–162, 1994.

104. L. Lee and J. L. Chen. Strictly positive real lemma and absolute stability for discrete-time descriptor systems. *IEEE Trans. Circuits Syst. I*, 50:788–794, 2003.

105. L. Lee and C.-H. Fang. An improved bound for stability robustness of uncertain generalized state-space systems. In *Proc. American Control Conference*, pages 240–241, Baltimore, Maryland, 1994.

106. L. Lee, C.-H. Fang, and J.-G. Hsieh. Exact unidirectional perturbation bounds for robustness of uncertain generalized state-space systems: continuous-time cases. *Automatica*, 33:1923–1927, 1997.

107. F. L. Lewis. Descriptor systems: Decomposition into forward and backward subsystems. *IEEE Trans. Automat. Control*, 29:167–170, 1984.

108. F. L. Lewis. A survey of linear singular systems. *Circuits, Syst. Signal Processing*, 5:3–36, 1986.

109. X. Li and C. E. de Souza. Criteria for robust stability and stabilization of uncertain linear systems with state-delay. *Automatica*, 33:1657–1662, 1997.

110. X. Li and C. E. de Souza. Delay-dependent robust stability and stabilization of uncertain linear delay systems: a linear matrix inequality approach. *IEEE Trans. Automat. Control*, 42:1144–1148, 1997.

111. C. Lin, J. Wang, and C. B. Soh. Maximum bounds for robust stability of linear uncertain descriptor systems with structured perturbations. *Int. J. Systems Sci.*, 34:463–467, 2003.

112. C. Lin, J. Wang, D. Wang, and C. B. Soh. Robustness of uncertain descriptor systems. *Systems & Control Lett.*, 31:129–138, 1997.

113. C. Lin, J. Wang, G.-H. Yang, and J. Lam. Robust stabilization via state feedback for descriptor systems with uncertainties in the derivative matrix. *Int. J. Control*, 73:407–415, 2000.

114. C. Lin, J. Wang, G.-H. Yang, and C. B. Soh. Robust C-controllability and/or C-observability for uncertain descriptor systems with interval perturbations in all matrices. *IEEE Trans. Automat. Control*, 44:1768–1773, 1999.

115. C. Lin, Q.-G. Wang, and T. H. Lee. Robust normalization and stabilization of uncertain descriptor systems with norm-bounded perturbations. *IEEE Trans. Automat. Control*, 50:515–520, 2005.

116. J. L. Lin and S. J. Chen. LFT approach to robust D-stability bounds of uncertain linear singular systems. *IEE Proc.-Control Theory Appl.*, 145:127–134, 1998.

117. W. Q. Liu, K. L. Teo, and W. Y. Yan. Optimal simultaneous stabilization of descriptor systems via output feedback. *Int. J. Control*, 64:595–613, 1996.

118. X. Liu. Local disturbance decoupling of nonlinear singular systems. *Int. J. Control*, 70:685–702, 1998.

119. X. Liu and S. Čelikovský. Feedback control of affine nonlinear singular control systems. *Int. J. Control*, 68:753–774, 1997.

120. A. Luenberger, D. G.and Arbel. Singular dynamic Leontief systems. *Econometrica*, 45:991–995, 1977.

121. D. G. Luenberger. Dynamic equations in descriptor form. *IEEE Trans. Automat. Control*, 22:312–321, 1977.

122. D. G. Luenberger. Time-invariant descriptor systems. *Automatica*, 14:473–480, 1978.

123. M. Mariton. *Jump Linear Systems in Automatic Control*. New York : Marcel Dekker, 1990.

124. B. Marx, D. Koenig, and D. Georges. Robust fault-tolerant control for descriptor systems. *IEEE Trans. Automat. Control*, 49:1869–1875, 2004.

125. I. Masubuchi, Y. Kamitane, A. Ohara, and N. Suda. H_∞ control for descriptor systems: a matrix inequalities approach. *Automatica*, 33:669–673, 1997.

126. J. K. Mills and A. A. Goldenberg. Force and position control of manipulators during constrained motion tasks. *IEEE Trans. Robot. Automat.*, 5:30–46, 1989.

127. N. Minamide, Y. Fujisaki, and A. Shimizu. A parametrization of all observers for descriptor systems. *Int. J. Control*, 66:767–777, 1997.

128. P. Misra and R. V. Patel. Computation of minimal-order realizations of generalized state-space systems. *Circuit, Syst, Sig. Process.*, 8:49–70, 1989.

129. P. C. Muller and M. Hou. On the observer design for descriptor systems. *IEEE Trans. Automat. Control*, 38:1666–1671, 1993.

130. K. M. Nagpal and P. P. Khargonekar. Filtering and smoothing in an H_∞ setting. *IEEE Trans. Automat. Control*, 36:152–166, 1991.

131. R. W. Newcomb. The semistate description of nonlinear time-variable circuits. *IEEE Trans. Circuits Syst.*, 28:62–71, 1981.

132. R. W. Newcomb and B. Dziurla. Some circuits and systems applications of semistate theory. *Circuit, Syst, Sig. Process.*, 8:235–260, 1989.

133. N. K. Nichols. On the stability radius of a generalized state-space system. *Linear Algebra Appl.*, 188, 1993.

134. S.-I. Niculescu. *Delay Effects on Stability : A Robust Control Approach*. Berlin : Springer, 2001.

135. R. Nikoukhah, S. L. Campbell, and F. Delebecque. Kalman filtering for general discrete-time linear systems. *IEEE Trans. Automat. Control*, 44:1829–1839, 1999.

136. K. Ozcaldiran and F. L. Lewis. On the regularizability of singular systems. *IEEE Trans. Automat. Control*, 35:1156–1160, 1990.

137. L. Pandolfi. Controllability and stabilization for linear systems of algebraic and differential equations. *J. Optim. Theory Appl.*, 30:601–620, 1980.

138. L. Pandolfi. On the regulator problem for linear degenerate control systems. *J. Optim. Theory Appl.*, 33:241–254, 1981.

139. S. Pang, J. Huang, and Y. Bai. Robust output regulation of singular nonlinear systems via a nonlinear internal model. *IEEE Trans. Automat. Control*, 50:222–228, 2005.

140. P. N. Paraskevopoulos and F. N. Koumboulis. Unifying approach to observers for regular and singular systems. *IEE Proc.-Control Theory Appl.*, 138:561–572, 1991.

141. P. N. Paraskevopoulos and F. N. Koumboulis. Observers for singular systems. *IEEE Trans. Automat. Control*, 37:1211–1215, 1992.

142. P. G. Park and T. Kailath. H_∞ filtering via convex optimization. *Int. J. Control*, 66:15–22, 1997.

143. I. R. Petersen and D. C. McFarlane. Optimal guaranteed cost control and filtering for uncertain linear systems. *IEEE Trans. Automat. Control*, 39:1971–1977, 1994.

144. L. Qiu and E. J. Davison. The stability robustness of generalized eigenvalues. *IEEE Trans. Automat. Control*, 37:886–891, 1992.

145. J. Rodriguez and D. Sweet. A characterization of semistate systems. *Circuit, Syst, Sig. Process.*, 5:125–137, 1986.

146. B. Scott. Power system dynamic response calculations. *Proc. IEEE*, 67:219–247, 1979.

147. U. Shaked and Y. Theodor. H_∞-optimal estimation: a tutorial. In *Proc. 31st IEEE Conf. Decision and Control*, pages 2278–2286, Tucson, Arizona, USA, December, 1992.

148. A. A. Shcheglova and V. F. Chistyakov. Stability of linear differential-algebraic systems. *Differ. Equ.*, 40:50–62, 2004.

149. P. Shi, E. K. Boukas, and R. K. Agarwal. Kalman filtering for continuous-time uncertain systems with Markovian jumping parameters. *IEEE Trans. Automat. Control*, 44:1592–1597, 1999.

150. D. N. Shields. Observer design for discrete nonlinear descriptor systems. *Int. J. Systems Sci.*, 27:1033–1041, 1996.

151. S. Singh and R.-W. Liu. Existence of state equation representation of linear large-scale dynamical systems. *IEEE Trans. Circuits Syst.*, 20:239–246, 1973.

152. B. L. Stevens and F. L. Lewis. *Aircraft Modeling, Dynamics and Control.* New York: Wiley, 1991.

153. W. Sun, P. P. Khargonekar, and D. Shim. Solution to the positive real control problem for linear time-invariant systems. *IEEE Trans. Automat. Control*, 39:2034–2046, 1994.

154. V. L. Syrmos, P. Misra, and R. Aripirala. On the discrete generalized Lyapunov equation. *Automatica*, 31:297–301, 1995.

155. K. Takaba and T. Katayama. H_2 output feedback control for descriptor systems. *Systems & Control Lett.*, 34:841–850, 1988.

156. K. Takaba and T. Katayama. Discrete-time H_∞ algebraic Riccati equation and parametrization of all H_∞ filters. *Int. J. Control*, 64:1129–1149, 1996.

157. K. Takaba, N. Morihira, and T. Katayama. H_∞ control for descriptor systems–a J-spectral factorization approach. In *Proc. 33rd IEEE Conf. Decision and Control*, pages 2251–2256, Lake Buena Vista, FL, USA, 1994.

158. K. Takaba, N. Morihira, and T. Katayama. A generalized Lyapunov theorem for descriptor system. *Systems & Control Lett.*, 24:49–51, 1995.

159. A. Tornambè. A simple procedure for the stabilization of a class of uncontrollable generalized systems. *IEEE Trans. Automat. Control*, 41:603–607, 1996.

160. H. D. Tuan and P. Apkarian. Low nonconvexity-rank bilinear matrix inequalities: algorithms and applications in robust controller and structure designs. *IEEE Trans. Automat. Control*, 45:2111–2117, 2000.

161. H. D. Tuan, P. Apkarian, and Y. Nakashima. A new lagrangian dual global optimization algorithm for solving bilinear matrix inequalities. *Int. J. Robust & Nonlinear Control*, 10:561–578, 2000.

162. D. Vafiadis and N. Karcanias. Canonical forms for descriptor systems under restricted system equivalence. *Automatica*, 33:955–958, 1997.

163. D. Vafiadis and N. Karcanias. Block decoupling and pole assignment of singular systems: frequency domain approach. *Int. J. Control*, 76:185–192, 2003.

164. A. Varga. On stabilization methods of descriptor systems. *Systems & Control Lett.*, 24:133–138, 1995.

165. G. C. Verghese, B. C. Lévy, and T. Kailath. A generalized state-space for singular systems. *IEEE Trans. Automat. Control*, 26:811–831, 1981.

166. E. I. Verriest and A. F. Ivanov. Robust stability of delay-difference equations. In *Proc. 34th IEEE Conf. Decision and Control*, pages 386–391, New Orleans, LA, December, 1995.

167. M. Vidyasagar. *Nonlinear Systems Analysis*. Prentice-Hall, Englewood Cliffs, New Jersey, 1993.

168. C. J. Wang. On linear quadratic optimal control of linear time-varying singular systems. *Int. J. Systems Sci.*, 35:903–906, 2004.

169. D. Wang and C. B. Soh. On regularizing singular systems by decentralized output feedback. *IEEE Trans. Automat. Control*, 44:148–152, 1999.

170. H.-S. Wang, C.-F. Yung, and F.-R. Chang. Bounded real lemma and H_∞ control for descriptor systems. *IEE Proc.-Control Theory Appl.*, 145:316–322, 1998.

171. Y.-Y. Wang, S.-J. Shi, and Z.-J. Zhang. Pole placement and compensator design of generalized systems. *Systems & Control Lett.*, 8:205–209, 1987.

172. G. Xie and L. Wang. Controllability of linear descriptor systems. *IEEE Trans. Circuits Syst. I*, 50:455–460, 2003.

173. L. Xie and C. E. de Souza. Robust H_∞ control for linear systems with norm-bounded time-varying uncertainty. *IEEE Trans. Automat. Control*, 37:1188–1191, 1992.

174. L. Xie and Y. C. Soh. Guaranteed cost control of uncertain discrete-time systems. *Control Theory Adv. Tech.*, 10:1235–1251, 1995.

175. L. Xie and Y. C. Soh. Positive real control for uncertain linear time-invariant systems. *Systems & Control Lett.*, 24:265–271, 1995.

176. S. Xu and J. Lam. H_∞ model reduction for discrete-time singular systems. *Systems & Control Lett.*, 48:121–133, 2003.

177. S. Xu and J. Lam. Robust stability for uncertain discrete singular systems with state delay. *Asian Journal of Control*, 5:399–405, 2003.

178. S. Xu and J. Lam. New positive realness conditions for uncertain discrete descriptor systems: analysis and synthesis. *IEEE Trans. Circuits Syst. I*, 51:1897–1905, 2004.

179. S. Xu and J. Lam. Robust stability and stabilization of discrete singular systems: an equivalent characterization. *IEEE Trans. Automat. Control*, 49, 2004.

180. S. Xu, J. Lam, W. Liu, and Q. L. Zhang. H_∞ model reduction for singular systems: continuous-time case. *IEE Proc.-Control Theory Appl.*, 150:637–641, 2003.

181. S. Xu, J. Lam, and C. Yang. H_∞ and positive real control for linear neutral delay systems. *IEEE Trans. Automat. Control*, 46:1321–1326, 2001.

182. S. Xu, J. Lam, and C. Yang. Quadratic stability and stabilization of uncertain linear discrete-time systems with state delay. *Systems & Control Lett.*, 43:77–84, 2001.

183. S. Xu, J. Lam, and C. Yang. Robust H_∞ control for discrete singular systems with state delay and parameter uncertainty. *Dynamics of Continuous, Discrete and Impulsive Systems, Series B: Applications and Algorithms*, 9:539–554, 2002.

184. S. Xu, J. Lam, and C. Yang. Robust H_∞ control for uncertain singular systems with state delay. *Int. J. Robust & Nonlinear Control*, 13:1213–1223, 2003.

185. S. Xu, J. Lam, C. Yang, and E. I. Verriest. An LMI approach to guaranteed cost control for uncertain linear neutral delay systems. *Int. J. Robust & Nonlinear Control*, 13:35–53, 2003.

186. S. Xu, J. Lam, and L. Zhang. Robust D-stability analysis for uncertain discrete singular systems with state delay. *IEEE Trans. Circuits Syst. I*, 49:551–555, 2002.

187. S. Xu, J. Lam, and Y. Zou. H_∞ filtering for singular systems. *IEEE Trans. Automat. Control*, 48:2217–2222, 2003.

188. S. Xu, P. Van Dooren, R. Stefan, and J. Lam. Robust stability and stabilization for singular systems with state delay and parameter uncertainty. *IEEE Trans. Automat. Control*, 47:1122–1128, 2002.

189. S. Xu and C. Yang. Stabilization of discrete-time singular systems: a matrix inequalities approach. *Automatica*, 35:1613–1617, 1999.

190. S. Xu and C. Yang. An algebraic approach to the robust stability analysis and robust stabilization of uncertain singular systems. *Int. J. Systems Sci.*, 31:55–61, 2000.

191. S. Xu and C. Yang. H_∞ state feedback control for discrete singular systems. *IEEE Trans. Automat. Control*, 45:1405–1409, 2000.

192. S. Xu, C. Yang, Y. Niu, and J. Lam. Robust stabilization for uncertain discrete singular systems. *Automatica*, 37:769–774, 2001.

193. I. Yaesh and U. Shaked. Game theory approach to optimal linear state estimation and its relation to the minimum H_∞-norm estimation. *IEEE Trans. Automat. Control*, 37:828–831, 1992.

194. C. W. Yang and H. L. Tan. Observer design for singular systems with unknown inputs. *Int. J. Control*, 49:1937–1946, 1989.

195. E. L. Yip and R. F. Sincovec. Solvability, controllability and observability of continuous descriptor systems. *IEEE Trans. Automat. Control*, 26:702–707, 1981.

196. L. Yu and J. Chu. An LMI approach to guaranteed cost control of linear uncertain time-delay systems. *Automatica*, 35:1155–1159, 1999.

197. R. Yu and D. Wang. Structural properties and poles assignability of LTI singular systems under output feedback. *Automatica*, 39:685–692, 2003.

198. D. Yue and Q.-L. Han. Robust H_∞ filter design of uncertain descriptor systems with discrete and distributed delays. *IEEE Trans. Signal Processing*, 52:3200–3212, 2004.

199. D. Yue and J. Lam. Suboptimal robust mixed H_2/H_∞ controller design for uncertain descriptor systems with distributed delays. *Comput. Math. Appl.*, 47:1041–1055, 2004.

200. G. Zhang, Q. L. Zhang, T. Chen, and Y. Lin. On Lyapunov theorems for descriptor systems. *Discrete Contin. Dyn. Syst.-Ser. B*, 10:709–725, 2003.

201. L. Zhang and B. Huang. Robust model predictive control of singular systems. *IEEE Trans. Automat. Control*, 49:1000–1011, 2004.

202. L. Zhang, B. Huang, and J. Lam. Robust guaranteed cost control of descriptor systems. *Discrete Contin. Dyn. Syst.-Ser. B*, 10:633–646, 2003.
203. L. Zhang, J. Lam, and S. Xu. On positive realness of descriptor systems. *IEEE Trans. Circuits Syst. I*, 49:401–407, 2002.
204. L. Zhang, J. Lam, and Q. L. Zhang. Lyapunov and Riccati equations of discrete-time descriptor systems. *IEEE Trans. Automat. Control*, 44:2134–2139, 1999.
205. L. Zhang, J. Lam, and Q. L. Zhang. Optimal model reduction of discrete-time descriptor systems. *Int. J. Systems Sci.*, 32:575–583, 2001.
206. Q. L. Zhang, J. Lam, and L. Zhang. On analyzing the stability of discrete descriptor systems via generalized Lyapunov equations. In *Multi-objective Programming and Goal Programming*, Adv. Soft Comput., pages 295–300. Springer, Berlin, 2003.
207. Q. L. Zhang, W. Q. Liu, and D. Hill. A Lyapunov approach to analysis of discrete singular systems. *Systems & Control Lett.*, 45:237–247, 2002.
208. S. Y. Zhang. Pole placement for singular systems. *Systems & Control Lett.*, 12:339–342, 1989.
209. K. Zhou, J. C. Doyle, and K. Glover. *Robust and Optimal Control*. Prentice-Hall, Upper Saddle River, NJ., 1996.
210. K. Zhou and P. P. Khargonekar. Robust stabilization of linear systems with norm-bounded time-varying uncertainty. *Systems & Control Lett.*, 10:17–20, 1988.
211. S. Zhou and J. Lam. Robust stabilization of delayed singular systems with linear fractional parametric uncertainties. *Circuit, Syst, Sig. Process.*, 22:579–588, 2003.
212. Z. Zhou, M. A. Shayman, and T.-J. Tarn. Singular systems: a new approach in the time domain. *IEEE Trans. Automat. Control*, 32:42–50, 1987.
213. J. Zhu, S. Ma, and Z. Cheng. Singular LQ problem for nonregular descriptor systems. *IEEE Trans. Automat. Control*, 47:1128–1133, 2002.
214. G. Zimmer and J. Meier. On observing nonlinear descriptor systems. *Systems & Control Lett.*, 32:43–48, 1997.
215. Y. Zou and C. Yang. Formulas for the distance between controllable and uncontrollable linear-systems. *Systems & Control Lett.*, 21:173–180, 1993.

Index

Lecture Notes in Control and Information Sciences

Edited by M. Thoma and M. Morari

Further volumes of this series can be found on our homepage:
springer.com

Vol. 305: Nebylov, A.
Ensuring Control Accuracy
256 p. 2004 [3-540-21876-9]

Vol. 304: Margaris, N.I.
Theory of the Non-linear Analog Phase Locked Loop
303 p. 2004 [3-540-21339-2]

Vol. 303: Mahmoud, M.S.
Resilient Control of Uncertain Dynamical Systems
278 p. 2004 [3-540-21351-1]

Vol. 302: Filatov, N.M.; Unbehauen, H.
Adaptive Dual Control: Theory and Applications
237 p. 2004 [3-540-21373-2]

Vol. 301: de Queiroz, M.; Malisoff, M.; Wolenski, P. (Eds.)
Optimal Control, Stabilization and Nonsmooth Analysis
373 p. 2004 [3-540-21330-9]

Vol. 300: Nakamura, M.; Goto, S.; Kyura, N.; Zhang, T.
Mechatronic Servo System Control
Problems in Industries and their Theoretical Solutions
212 p. 2004 [3-540-21096-2]

Vol. 299: Tarn, T.-J.; Chen, S.-B.; Zhou, C. (Eds.)
Robotic Welding, Intelligence and Automation
214 p. 2004 [3-540-20804-6]

Vol. 298: Choi, Y.; Chung, W.K.
PID Trajectory Tracking Control for Mechanical Systems
127 p. 2004 [3-540-20567-5]

Vol. 297: Damm, T.
Rational Matrix Equations in Stochastic Control
219 p. 2004 [3-540-20516-0]

Vol. 296: Matsuo, T.; Hasegawa, Y.
Realization Theory of Discrete-Time Dynamical Systems
235 p. 2003 [3-540-40675-1]

Vol. 295: Kang, W.; Xiao, M.; Borges, C. (Eds)
New Trends in Nonlinear Dynamics and Control,
and their Applications
365 p. 2003 [3-540-10474-0]

Vol. 294: Benvenuti, L.; De Santis, A.; Farina, L. (Eds)
Positive Systems: Theory and Applications (POSTA 2003)
414 p. 2003 [3-540-40342-6]

Vol. 293: Chen, G. and Hill, D.J.
Bifurcation Control
320 p. 2003 [3-540-40341-8]

Vol. 292: Chen, G. and Yu, X.
Chaos Control
380 p. 2003 [3-540-40405-8]

Vol. 291: Xu, J.-X. and Tan, Y.
Linear and Nonlinear Iterative Learning Control
189 p. 2003 [3-540-40173-3]

Vol. 290: Borrelli, F.
Constrained Optimal Control
of Linear and Hybrid Systems
237 p. 2003 [3-540-00257-X]

Vol. 289: Giarré, L. and Bamieh, B.
Multidisciplinary Research in Control
237 p. 2003 [3-540-00917-5]

Vol. 288: Taware, A. and Tao, G.
Control of Sandwich Nonlinear Systems
393 p. 2003 [3-540-44115-8]

Vol. 287: Mahmoud, M.M.; Jiang, J.; Zhang, Y.
Active Fault Tolerant Control Systems
239 p. 2003 [3-540-00318-5]

Vol. 286: Rantzer, A. and Byrnes C.I. (Eds)
Directions in Mathematical Systems
Theory and Optimization
399 p. 2003 [3-540-00065-8]

Vol. 285: Wang, Q.-G.
Decoupling Control
373 p. 2003 [3-540-44128-X]

Vol. 284: Johansson, M.
Piecewise Linear Control Systems
216 p. 2003 [3-540-44124-7]

Vol. 283: Fielding, Ch. et al. (Eds)
Advanced Techniques for Clearance of
Flight Control Laws
480 p. 2003 [3-540-44054-2]

Vol. 282: Schröder, J.
Modelling, State Observation and
Diagnosis of Quantised Systems
368 p. 2003 [3-540-44075-5]

Vol. 281: Zinober A.; Owens D. (Eds)
Nonlinear and Adaptive Control
416 p. 2002 [3-540-43240-X]

Vol. 280: Pasik-Duncan, B. (Ed)
Stochastic Theory and Control
564 p. 2002 [3-540-43777-0]

Vol. 279: Engell, S.; Frehse, G.; Schnieder, E. (Eds)
Modelling, Analysis, and Design of Hybrid Systems
516 p. 2002 [3-540-43812-2]

Vol. 278: Chunling D. and Lihua X. (Eds)
H_∞ Control and Filtering of
Two-dimensional Systems
161 p. 2002 [3-540-43329-5]

Vol. 277: Sasane, A.
Hankel Norm Approximation
for Infinite-Dimensional Systems
150 p. 2002 [3-540-43327-9]

Vol. 276: Bubnicki, Z.
Uncertain Logics, Variables and Systems
142 p. 2002 [3-540-43235-3]

Vol. 275: Ishii, H.; Francis, B.A.
Limited Data Rate in Control Systems with Networks
171 p. 2002 [3-540-43237-X]